Advanced Design Problems in Aerospace Engineering

Volume 1: Advanced Aerospace Systems

MATHEMATICAL CONCEPTS AND METHODS IN SCIENCE AND ENGINEERING

Series Editor: **Angelo Miele**
George R. Brown School of Engineering
Rice University

Recent volumes in this series:

31 **NUMERICAL DERIVATIVES AND NONLINEAR ANALYSIS**
• *Harriet Kagiwada, Robert Kalaba, Nima Rasakhoo, and Karl Spingarn*

32 **PRINCIPLES OF ENGINEERING MECHANICS**
Volume 1: Kinematics—The Geometry of Motion • *M. F. Beatty, Jr.*

33 **PRINCIPLES OF ENGINEERING MECHANICS**
Volume 2: Dynamics—The Analysis of Motion • *Millard F. Beatty, Jr.*

34 **STRUCTURAL OPTIMIZATION**
Volume 1: Optimality Criteria • *Edited by M. Save and W. Prager*

35 **OPTIMAL CONTROL APPLICATIONS IN ELECTRIC POWER SYSTEMS**
• *G. S. Christensen, M. E. El-Hawary, and S. A. Soliman*

36 **GENERALIZED CONCAVITY**
• *Mordecai Avriel, Walter W. Diewert, Siegfried Schaible, and Israel Zang*

37 **MULTICRITERIA OPTIMIZATION IN ENGINEERING AND IN THE SCIENCES**
• *Edited by Wolfram Stadler*

38 **OPTIMAL LONG-TERM OPERATION OF ELECTRIC POWER SYSTEMS**
• *G. S. Christensen and S. A. Soliman*

39 **INTRODUCTION TO CONTINUUM MECHANICS FOR ENGINEERS**
• *Ray M. Bowen*

40 **STRUCTURAL OPTIMIZATION**
Volume 2: Mathematical Programming • *Edited by M. Save and W. Prager*

41 **OPTIMAL CONTROL OF DISTRIBUTED NUCLEAR REACTORS**
• *G. S. Christensen, S. A. Soliman, and R. Nieva*

42 **NUMERICAL SOLUTIONS OF INTEGRAL EQUATIONS**
• *Edited by Michael A. Golberg*

43 **APPLIED OPTIMAL CONTROL THEORY OF DISTRIBUTED SYSTEMS**
• *K. A. Lurie*

44 **APPLIED MATHEMATICS IN AEROSPACE SCIENCE AND ENGINEERING**
• *Edited by Angelo Miele and Attilio Salvetti*

45 **NONLINEAR EFFECTS IN FLUIDS AND SOLIDS**
• *Edited by Michael M. Carroll and Michael A. Hayes*

46 **THEORY AND APPLICATIONS OF PARTIAL DIFFERENTIAL EQUATIONS**
• *Piero Bassanini and Alan R. Elcrat*

47 **UNIFIED PLASTICITY FOR ENGINEERING APPLICATIONS**
• *Sol R. Bodner*

48 **ADVANCED DESIGN PROBLEMS IN AEROSPACE ENGINEERING**
Volume 1: Advanced Aerospace Systems • *Edited by Angelo Miele and Aldo Frediani*

Advanced Design Problems in Aerospace Engineering

Volume 1: Advanced Aerospace Systems

Edited by

Angelo Miele

Rice University
Houston, Texas

and

Aldo Frediani

University of Pisa
Pisa, Italy

Springer Science+Business Media, LLC

ISBN 978-1-4419-3448-2 ISBN 978-0-306-48637-1 (eBook)
DOI 10.1007/978-0-306-48637-1

Originally published by Kluwer Academic / Plenum Publishers in 2003
Softcover reprint of the hardcover 1st edition 2003

http://www.wkap.nl/

10 9 8 7 6 5 4 3 2 1

A C.I.P. record for this book is available from the Library of Congress

Contributors

P. Alli, Agusta Corporation, 21017 Cascina di Samarate, Varese, Italy.

G. Bernardini, Department of Mechanical and Industrial Engineering, University of Rome-3, 00146 Rome, Italy.

A. Beukers, Faculty of Aerospace Engineering, Delft University of Technology, 2629 HS Delft, Netherlands.

V. Caramaschi, Agusta Corporation, 21017 Cascina di Samarate, Varese, Italy.

M. Chiarelli, Department of Aerospace Engineering, University of Pisa, 56100 Pisa, Italy.

T. De Jong, Faculty of Aerospace Engineering, Delft University of Technology, 2629 HS Delft, Netherlands.

I. P. Fielding, Aerospace Design Group, Cranfield College of Aeronautics, Cranfield University, Cranfield, Bedforshire MK43 OAL, England.

A. Frediani, Department of Aerospace Engineering, University of Pisa, 56100 Pisa, Italy.

M. Hanel, Institute of Flight Mechanics and Flight Control, University of Stuttgart, 70550 Stuttgart, Germany.

J. Hinrichsen, Airbus Industries, 1 Round Point Maurice Bellonte, 31707 Blagnac, France.

L. A. Krakers, Faculty of Aerospace Engineering, Delft University of Technology, 2629 HS Delft, Netherlands.

A. Longhi, Department of Aerospace Engineering, University of Pisa, 56100 Pisa, Italy.

S. Mancuso, ESA-ESTEC Laboratory, 2201 AZ Nordwijk, Netherlands.

A. Miele, Aero-Astronautics Group, Rice University, Houston, Texas 77005-1892, USA.

G. Montanari, Department of Aerospace Engineering, University of Pisa, 56100 Pisa, Italy.

L. Morino, Department of Mechanical and Industrial Engineering, University of Rome-3, 00146 Rome, Italy.

F. Nannoni, Agusta Corporation, 21017 Cascina di Samarate, Varese, Italy.

M. Raggi, Agusta Corporation, 21017 Cascina di Samarate, Varese, Italy.

J. Roskam, DAR Corporation, 120 East 9th Street, Lawrence, Kansas 66044, USA.

G. Sachs, Institute of Flight Mechanics and Flight Control, Technical University of Munich, 85747 Garching, Germany.

H. Smith, Aerospace Design Group, Cranfield College of Aeronautics, Cranfield University, Cranfield, Bedforshire MK43 OAL, England.

E. Troiani, Department of Aerospace Engineering, University of Pisa, 56100 Pisa, Italy.

M.J.L. Van Tooren, Faculty of Aerospace Engineering, Delft University of Technology, 2629 HS Delft, Netherlands.

T. Wang, Aero-Astronautics Group, Rice University, Houston, Texas 77005-1892, USA.

K.H. Well, Institute of Flight Mechanics and Flight Control, University of Stuttgart, 70550 Stuttgart, Germany.

Preface

The meeting on "Advanced Design Problems in Aerospace Engineering" was held in Erice, Sicily, Italy from July 11 to July 18, 1999. The occasion of the meeting was the 28th Workshop of the School of Mathematics "Guido Stampacchia", directed by Professor Franco Giannessi of the University of Pisa. The School is affiliated with the International Center for Scientific Culture "Ettore Majorana", which is directed by Professor Antonino Zichichi of the University of Bologna.

The intent of the Workshop was the presentation of a series of lectures on the use of mathematics in the conceptual design of various types of aircraft and spacecraft. Both atmospheric flight vehicles and space flight vehicles were discussed. There were 16 contributions, six dealing with Advanced Aerospace Systems and ten dealing with Unconventional and Advanced Aircraft Design. Accordingly, the proceedings are split into two volumes.

The first volume (this volume) covers topics in the areas of flight mechanics and astrodynamics pertaining to the design of Advanced Aerospace Systems. The second volume covers topics in the areas of aerodynamics and structures pertaining to Unconventional and Advanced Aircraft Design. An outline is given below.

Advanced Aerospace Systems

Chapter 1, by A. Miele and S. Mancuso (Rice University and ESA/ESTEC), deals with the design of rocket-powered orbital spacecraft. Single-stage configurations are compared with double-stage configurations using the sequential gradient-restoration algorithm in optimal control format.

Chapter 2, by A. Miele and S. Mancuso (Rice University and ESA/ESTEC), deals with the design of Moon missions. Optimal outgoing and return trajectories are determined using the sequential gradient-restoration algorithm in mathematical programming format. The analyses are made within the frame of the restricted three-body problem and the results are interpreted in light of the theorem of image trajectories in Earth-Moon space.

Chapter 3, by A. Miele and T. Wang (Rice University), deals with the design of Mars missions. Optimal outgoing and return trajectories are determined using the sequential gradient-restoration algorithm in mathematical programming format. The analyses are made within the frame of the restricted four-body problem and the results are interpreted in light of the relations between outgoing and return trajectories.

Chapter 4, by G. Sachs (Technical University of Munich), deals with the design and test of an experimental guidance system with perspective flight path display. It considers the design issues of a guidance system displaying visual information to the pilot in a three-dimensional format intended to improve manual flight path control.

Chapter 5, by K.H. Well (University of Stuttgart), deals with the neighboring vehicle design for a two-stage launch vehicle. It is concerned with the optimization of the ascent trajectory of a two-stage launch vehicle simultaneously with the optimization of some significant design parameters.

Chapter 6, by M. Hanel and K.H. Well (University of Stuttgart), deals with the controller design for a flexible aircraft. It presents an overview of the models governing the dynamic behavior of a large four-engine flexible aircraft. It considers several alternative options for control system design.

Unconventional Aircraft Design

Chapter 7, by J.P. Fielding and H. Smith (Cranfield College of Aeronautics), deals with conceptual and preliminary methods for use on conventional and blended wing-body airliners. Traditional design methods have concentrated largely on aerodynamic techniques, with some allowance made for structures and systems. New multidisciplinary design tools are developed and examples are shown of ways and means useful for tradeoff studies during the early design stages.

Chapter 8, by A. Frediani and G. Montanari (University of Pisa), deals with the Prandtl best-wing system. It analyzes the induced drag of a lifting multiwing system. This is followed by a box-wing system and then by the Prandtl best-wing system.

Chapter 9, by A. Frediani, A. Longhi, M. Chiarelli, and E. Troiani (University of Pisa), deals with new large aircraft with nonconventional configuration. This design is called the Prandtl plane and is a biplane with twin horizontal and twin vertical swept wings. Its induced drag is smaller than that of any aircraft with the same dimensions. Its structural, aerodynamic, and aeroelastic properties are discussed.

Chapter 10, by L. Morino and G. Bernardini (University of Rome-3), deals with the modeling of innovative configurations using multidisciplinary optimization (MDO) in combination with recent aerodynamic developments. It presents an overview of the techniques for modeling the structural, aerodynamic, and aeroelastic properties of aircraft, to be used in the preliminary design of innovative configurations via multidisciplinary optimization.

Advanced Aircraft Design

Chapter 11, by P. Alli, M. Raggi, F. Nannoni, and V. Caramaschi (Agusta Corporation), deals with the design problems for new helicopters. These problems are treated in light of the following aspects: man-machine interface, fly-by-wire, rotor aerodynamics, rotor dynamics, aeroelasticity, and noise reduction.

Chapter 12, by A. Beukers, M.J.L Van Tooren, and T. De Jong (Delft University of Technology), deals with a multidisciplinary design philosophy for aircraft fuselages. It treats the combined development of new materials, structural concepts, and manufacturing technologies leading to the fulfillment of appropriate mechanical requirements and ease of production.

Chapter 13, by A. Beukers, M.J.L. Van Tooren, T. De Jong, and L.A. Krakers (Delft University of Technology), continues Chapter 12 and deals with examples illustrating the multidisciplinary concept. It discusses the following problems: (a) tension-loaded plate with stress concentrations, (b) buckling of a composite plate, and (c) integration of acoustics and aerodynamics in a stiffened shell fuselage.

Chapter 14, by J. Hinrichsen (Airbus Industries), deals with the design features and structural technologies for the family of Airbus A3XX aircraft. It reviews the problems arising in the development of the A3XX aircraft family with respect to configuration design, structural design, and application of new materials and manufacturing technologies.

Chapter 15, by J. Roskam (DAR Corporation), deals with user-friendly general aviation airplanes via a revolutionary but affordable approach. It discusses the development of personal transportation airplanes as worldwide standard business tools. The areas covered include system design and integration as well as manufacturing at an acceptable cost level.

Chapter 16, by J. Roskam (DAR Corporation), deals with the design of a 10-20 passenger jet-powered regional transport and resulting economic challenges. It discusses the introduction of new small passenger jet aircraft designed for short-to-medium ranges. It proposes the development of a family of two airplanes: a single-fuselage 10-passenger airplane and a twin-fuselage 20-passenger airplane.

In closing, the Workshop Directors express their thanks to Professors Franco Giannessi and Antonino Zichichi for their contributions.

A. Miele
Rice University
Houston, Texas, USA

A. Frediani
University of Pisa
Pisa, Italy

Contents

1

Design of Rocket-Powered Orbital Spacecraft[1]

A. MIELE[2] AND S. MANCUSO[3]

Abstract. In this paper, the feasibility of single-stage-suborbital (SSSO), single-stage-to-orbit (SSTO), and two-stage-to-orbit (TSTO) rocket-powered spacecraft is investigated using optimal control theory. Ascent trajectories are optimized for different combinations of spacecraft structural factor and engine specific impulse, the optimization criterion being the maximum payload weight. Normalized payload weights are computed and used to assess feasibility.

The results show that SSSO feasibility does not necessarily imply SSTO feasibility: while SSSO feasibility is guaranteed for all the parameter combinations considered, SSTO feasibility is guaranteed for only certain parameter combinations, which might be beyond the present state of the art. On the other hand, not only TSTO feasibility is guaranteed for all the parameter combinations considered, but a TSTO spacecraft is considerably superior to a SSTO spacecraft in terms of payload weight.

Three areas of potential improvements are discussed: (i) use of lighter materials (lower structural factor) has a significant effect on payload weight and feasibility; (ii) use of engines with higher ratio of thrust to propellant weight flow (higher specific impulse) has also

---

[1] This paper is based on Refs. 1-4.

[2] Research Professor and Foyt Professor Emeritus of Engineering, Aerospace Sciences, and Mathematical Sciences, Aero-Astronautics Group, Rice University, Houston, Texas 77005-1892, USA.

[3] Guidance, Navigation, and Control Engineer, European Space Technology and Research Center, 2201 AZ, Nordwijk, Netherlands.

a significant effect on payload weight and feasibility; (iii) on the other hand, aerodynamic improvements via drag reduction have a relatively minor effect on payload weight and feasibility.

In light of (i) to (iii), with reference to the specific impulse/structural factor domain, nearly-universal zero-payload lines can be constructed separating the feasibility region (positive payload) from the unfeasibility region (negative payload). The zero-payload lines are of considerable help to the designer in assessing the feasibility of a given spacecraft.

Key Words. Flight mechanics, rocket-powered spacecraft, suborbital spacecraft, orbital spacecraft, optimal trajectories, ascent trajectories.

1. Introduction

After more than thirty years of development of multi-stage-to-orbit (MSTO) spacecraft, exemplified by the Space Shuttle and Ariane three-stage spacecraft, the natural continuation for a modern space program is the development of two-stage-to-orbit (TSTO) and then single-stage-to-orbit (SSTO) spacecraft (Refs. 1-7). The first step toward the latter goal is the development of a single-stage-suborbital (SSSO) rocket-powered spacecraft which must take-off vertically, reach given suborbital altitude and speed, and then land horizontally.

Within the above frame, this paper investigates via optimal control theory the feasibility of three different configurations: a SSSO configuration, exemplified by the X-33 spacecraft; a SSTO configuration, exemplified by the Venture Star spacecraft; and a TSTO configuration. Ascent trajectories are optimized for different combinations of spacecraft structural factor and engine specific impulse, the optimization criterion being the maximum payload weight. Realistic constraints are imposed on tangential acceleration, dynamic pressure, and heating rate.

The optimization is done employing the sequential gradient-restoration algorithm for optimal control problems (SGRA, Refs. 8-10), developed and perfected by the Aero-Astronautics Group of Rice University over the years. SGRA has the major property of being a robust algorithm, and it has been employed with success to solve a wide variety of aerospace problems (Refs. 11-16) including interplanetary trajectories (Ref. 11), flight in windshear (Refs. 12-13), aerospace plane trajectories (Ref. 14), and aeroassisted orbital transfer (Refs. 15-16).

In Section 2, we present the system description. In Section 3, we formulate the optimization problem and give results for the SSSO configuration. In Section 4, we formulate the optimization problem and give results for the SSTO configuration. In Sections 5, we formulate the optimization problem and give results for the TSTO configuration. Section 6 contains design considerations pointing out the areas of potential improvements. Finally, Section 7 contains the conclusions.

2. System Description

For all the configurations being studied, the following assumptions are employed: (A1) the flight takes place in a vertical plane over a spherical Earth; (A2) the Earth rotation is neglected; (A3) the gravitational field is central and obeys the inverse square law; (A4) the thrust is directed along the spacecraft reference line; hence, the thrust angle of attack is the same as the aerodynamic angle of attack; (A5) the spacecraft is controlled via the angle of attack and power setting.

2.1. Mathematical Model. With the above assumptions, the motion of the spacecraft is described by the following differential system for the altitude h, velocity V, flight path angle γ, and reference weight W (Ref. 17):

$$\dot{h} = V \sin\gamma, \tag{1a}$$

$$\dot{V} = (T/W)g_e \cos\alpha - (D/W)g_e - g\sin\gamma, \tag{1b}$$

$$\dot{\gamma} = (T/WV)g_e \sin\alpha + (L/WV)g_e + (V/r - g/V)\cos\gamma, \tag{1c}$$

$$W = -T/I_{sp}, \tag{1d}$$

in which the dot denotes derivative with respect to the time t. Here,

$0 \le t \le \tau$, where τ is the final time. The quantities on the right-hand side of (1) are the thrust T, drag D, lift L, reference weight W, radial distance r, local acceleration of gravity g, sea-level acceleration of gravity g_e, angle of attack α, and engine specific impulse I_{sp}.

In addition, the following relations hold:

$$r = r_e + h, \tag{2a}$$

$$g = \mu / r^2 = \mu / (r_e + h)^2, \tag{2b}$$

$$I_{sp} = V_{ex} / g_e, \tag{2c}$$

$$W = mg_e, \tag{2d}$$

where r_e is the Earth radius, $\mu = g_e r_e^2$ the Earth gravitational constant, V_{ex} the exit velocity of the gases, and m the instantaneous mass. Note that, by definition, the reference weight is proportional to the instantaneous mass.

The aerodynamic forces are given by

$$D = (1/2)C_D(\alpha, M)\rho(h)SV^2, \tag{3a}$$

$$L = (1/2)C_L(\alpha, M)\rho(h)SV^2, \tag{3b}$$

where C_D is the drag coefficient, C_L the lift coefficient, S a reference surface area, and ρ the air density (Ref. 18). Disregarding the dependence on the Reynolds number, the aerodynamic coefficients can be represented in terms of the angle of attack α and the Mach number $M = V/a(h)$, where a is the speed of sound. The functions $C_D(\alpha, M)$ and $C_L(\alpha, M)$ used in this paper are described in Refs. 1-4.

For the rocket powerplant under consideration, the following expressions are assumed for the thrust and specific impulse:

$$T = \beta T_*, \tag{4a}$$

$$I_{sp} = I_{sp*}, \tag{4b}$$

where β is the power setting, T_* a reference thrust (thrust for $\beta = 1$), and

I_{sp*} a reference specific impulse. The fact that T_* and I_{sp*} are assumed to be constant means that the weak dependence of T and I_{sp} on altitude and Mach number, relevant to a precision study, is disregarded within the present feasibility study.

The atmospheric model used is the 1976 US Standard Atmosphere (Ref. 18). In this model, the values of the density are tabulated at discrete altitudes. For intermediate altitudes, the density is computed by assuming an exponential fit for the function $\rho(h)$. This is equivalent to assuming that the atmosphere behaves isothermally between any two contiguous altitudes tabulated in Ref. 18.

2.2. Inequality Constraints. Inspection of the system (1) in light of (2)-(4) shows that the time history of the state $h(t)$, $V(t)$, $\gamma(t)$, $W(t)$ can be computed by forward integration for given initial conditions, given controls $\alpha(t)$ and $\beta(t)$, and given final time τ. In turn, the controls are subject to the two-sided inequality constraints

$$-3 \le \alpha \le +3 \text{ deg}, \tag{5a}$$

$$0 \le \beta \le 1, \tag{5b}$$

which must be satisfied everywhere along the interval of integration. In addition, some path constraints are imposed on tangential acceleration a_T, dynamic pressure q, and heating rate Q per unit time and unit surface area, specifically,

$$a_T = (T/W)g_e \cos\alpha - (D/W)g_e - g\sin\gamma, \qquad a_T \le 3g_e, \tag{6a}$$

$$q = (1/2)\rho(h)V^2, \qquad q \le q_{max}, \tag{6b}$$

$$Q = C\sqrt{\quad}[\rho(h)/\rho(h_*)](V/V_*)^{3.07}, \qquad Q \le Q_{max}. \tag{6c}$$

Note that (6a) involves directly both the state and the control; on the other hand, (6b) and (6c) involve directly the state and indirectly the control. Concerning (6c), h_* is a reference altitude, V_* is a reference velocity, and C is a dimensional constant; for details, see Refs. 1-4.

In solving the optimization problems, the control constraints (5) are accounted for via trigonometric transformations. On the other hand, the path constraints (6) are taken into account via penalty functionals.

2.3. Supplementary Data. The following data have been used in the numerical experiments:

radius of Earth $r_e = 6378$ km, (7a)

Earth gravitational constant $\mu = 3.986$ E+05 km^3/s^2, (7b)

sea-level acceleration of gravity $g_e = 9.807$ m/s^2. (7c)

3. Single-Stage Suborbital Spacecraft

The following data were considered for SSSO configurations designed to achieve Mach number $M = 15$ in level flight at $h = 76.2$ km:

initial wing loading $W_i / S = 854$ kgf/m^2, (8a)

initial thrust-to-weight ratio $T_i/W_i = 1.5$. (8b)

The values (8) are representative of the X-33 spacecraft.

3.1. Boundary Conditions. The initial conditions ($t = 0$, subscript i) and final conditions ($t = \tau =$ free, subscript f) are

$h_i = 0$ km, $h_f = 76.2$ km, (9a)

$V_i = 0$ km/s, $V_f = 4.307$ km/s, (9b)

$\gamma_i = \pi/2$, $\gamma_f = 0$, (9c)

$W_i/W_r = 1$, $W_f/W_r =$ free. (9d)

In Eqs. (9d), the reference weight W_r is the same as the takeoff weight.

3.2. Weight Distribution. The propellant weight W_p, structural weight W_s, and payload weight W_* can be expressed in terms of the initial weight W_i, final weight W_f, and structural factor ε via the following relations (Ref. 17):

$$W_p = W_i - W_f, \tag{10a}$$

$$W_s = \varepsilon (W_i - W_f) / (1 - \varepsilon), \tag{10b}$$

$$W_* = (W_f - \varepsilon W_i) / (1 - \varepsilon), \tag{10c}$$

with

$$\varepsilon = W_s / (W_s + W_p), \tag{10d}$$

$$W_i = W_p + W_s + W_* . \tag{10e}$$

3.3. Optimization Problem. For the SSSO configuration, the maximum payload problem can be formulated as follows [see (10c)]:

$$\max \quad W_* = W_f - W_s = (W_f - \varepsilon W_i) / (1 - \varepsilon), \tag{11a}$$

$$\text{s. t.} \quad (1)\text{-}(6) \text{ and } (9). \tag{11b}$$

The unknowns include the state variables h, V, γ, W, control variables α, β, and parameter τ.

3.4. Computer Runs. First, the maximum payload weight problem (11) was solved via the sequential gradient-restoration algorithm (SGRA) for the following combinations of engine specific impulse and spacecraft structural factor:

$$I_{sp} = 400, 420, 440, 460, 480, 500 \text{ s}, \tag{12a}$$

$$\varepsilon = 0.10, 0.16, 0.22. \tag{12b}$$

The results for the normalized final weight W_f/W_i , propellant weight W_p/W_i , structural weight W_s/W_i , and payload weight W_*/W_i associated with various parameter combinations can be found in Refs. 1 and 4. In Fig. 1a, the maximum value of the normalized payload weight is plotted versus the specific impulse for the values (12b) of the structural factor. The main comments are that:

(i) The normalized payload weight increases as the engine specific impulse increases and as the spacecraft structural factor decreases.
(ii) The design of the SSSO configuration is feasible for each of the parameter combinations (12).

Zero-Payload Line. Next assume that, for a given specific impulse in the range (12a), the structural factor is increased beyond the range (12b).

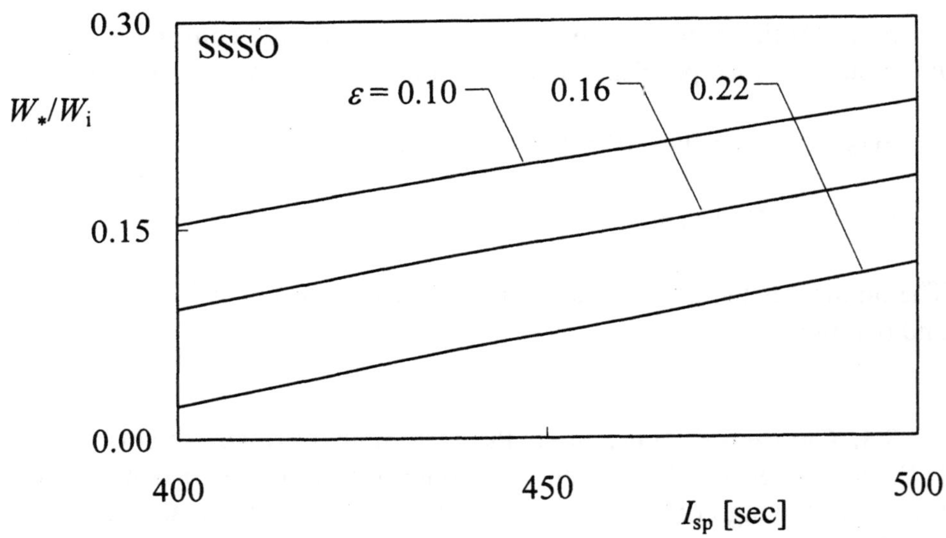

Fig. 1a. SSSO spacecraft: Normalized payload weight vs specific impulse and structural factor.

Each increase of ε causes a corresponding decrease in payload weight, until a limiting value ε_{max} is found such that $W_* = 0$. By repeating this procedure for each specific impulse in the range (12a), it is possible to construct a zero-payload line separating the feasibility region (below) from the unfeasibility region (above); this is shown in Fig. 1b with reference to the specific impulse/structural factor domain. The main comments are that:

(iii) Not only the zero-payload line supplies the upper bound ε_{max} ensuring feasibility for each given I_{sp}, but simultaneously supplies the lower bound $I_{sp\,min}$ ensuring feasibility for each given ε.

(iv) For a spacecraft of the X-33 type, with $I_{sp} = 420$ s, the limiting value of the structural factor is $\varepsilon_{max} = 0.255$. Should the SSSO design be such that $\varepsilon > 0.255$, it would become impossible for the X-33 spacecraft to reach the desired final Mach number $M_f = 15$ in level flight at the given final altitude $h_f = 76.2$ km. Instead, the spacecraft would reach a lower final Mach number, implying a subsequent decrease in range.

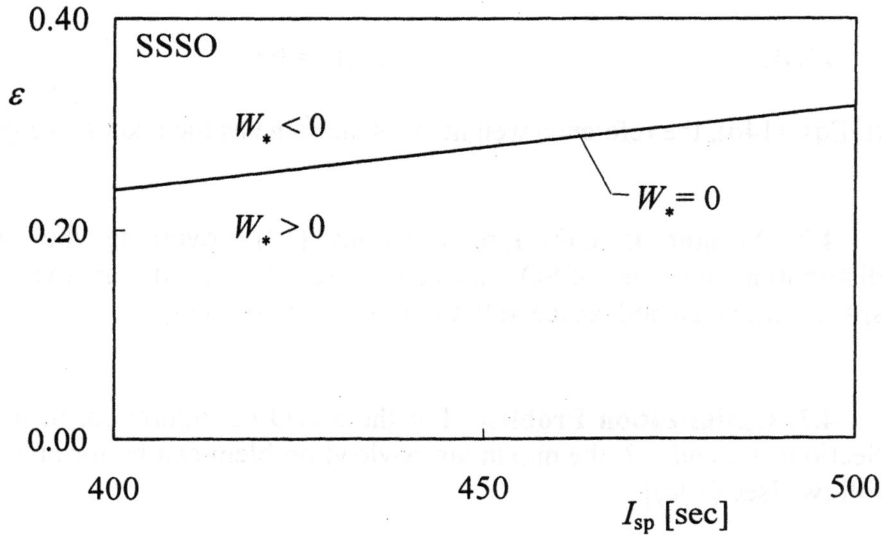

Fig. 1b. SSSO spacecraft, zero-payload line: Structural factor vs specific impulse.

4. Single-Stage Orbital Spacecraft

The following data were considered for SSTO configurations designed to achieve orbital speed at Space Station altitude, hence $V = 7.633$ km/s at $h = 463$ km:

initial wing loading $W_i / S = 1890$ kgf/m^2, (13a)

initial thrust-to-weight ratio $T_i/W_i = 1.4$. (13b)

The values (13) are representative of the Venture Star spacecraft.

4.1. Boundary Conditions. The initial conditions ($t = 0$, subscript i) and final conditions ($t = \tau =$ free, subscript f) are

$h_i = 0$ km, $h_f = 463$ km, (14a)

$V_i = 0$ km/s, $V_f = 7.633$ km/s, (14b)

$\gamma_i = \pi/2$, $\gamma_f = 0$, (14c)

$W_i/W_r = 1$, $W_f/W_r =$ free. (14d)

In Eqs. (14d), the reference weight W_r is the same as the takeoff weight.

4.2. Weight Distribution. Relations (10) governing the weight distribution for the SSSO spacecraft are also valid for the SSTO spacecraft, since both spacecraft are of the single-stage type.

4.3. Optimization Problem. For the SSTO configuration, in light of Sections 3.2 and 4.2, the maximum payload problem can be formulated as follows [see (10c)]:

max $W_* = W_f - W_s = (W_f - \varepsilon W_i) / (1 - \varepsilon)$, (15a)

s. t. (1)-(6) and (14). (15b)

The unknowns include the state variables h, V, γ, W, control variables α, β, and parameter τ.

4.4. Computer Runs. First, the maximum payload weight problem (15) was solved via SGRA for the following combinations of engine specific impulse and spacecraft structural factor:

$$I_{sp} = 400, 420, 440, 460, 480, 500 \text{ s}, \tag{16a}$$

$$\varepsilon = 0.08, 0.10, 0.12. \tag{16b}$$

The results for the normalized final weight W_f/W_i, propellant weight W_p/W_i, structural weight W_s/W_i, and payload weight W_*/W_i associated with various parameter combinations can be found in Refs. 2 and 4. In Fig. 2a, the maximum value of the normalized payload weight is plotted versus

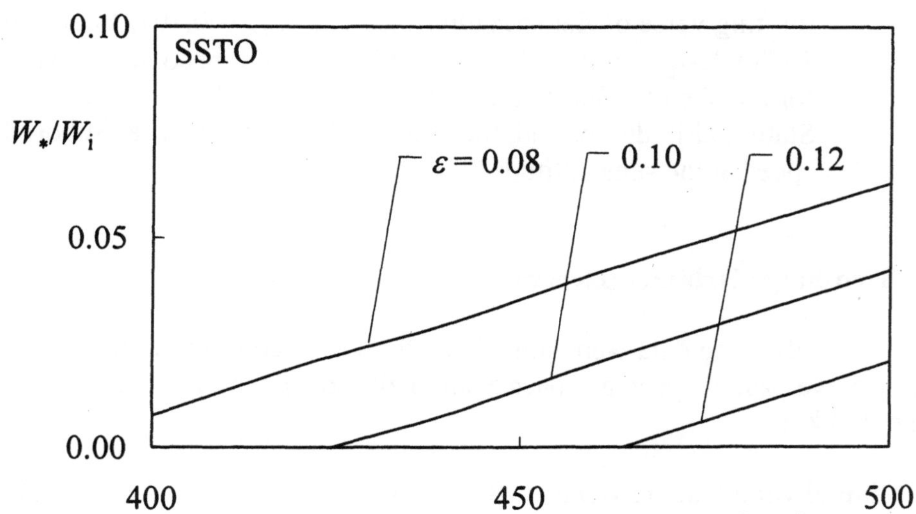

Fig. 2a. SSTO spacecraft: Normalized payload weight vs specific impulse and structural factor.

the specific impulse for the values (16b) of the structural factor. The main comments are that:

(i) The normalized payload weight increases as the engine specific impulse increases and as the spacecraft structural factor decreases.

(ii) The design of SSTO configurations might be comfortably feasible, marginally feasible, or unfeasible, depending on the parameter values assumed.

Zero-Payload Line. By proceeding along the lines of Section 3.4, a zero-payload line $W_* = 0$ can be constructed for the SSTO spacecraft. With reference to the specific impulse/structural factor domain, the zero-payload line is shown in Fig. 2b and separates the feasibility region (below) from the unfeasibility region (above). The main comments are that:

(iii) Not only the zero-payload line supplies the upper bound ε_{max} ensuring feasibility for each given I_{sp}, but simultaneously supplies the lower bound $I_{sp\,min}$ ensuring feasibility for each given ε.

(iv) For a spacecraft of the Venture Star type, with $I_{sp} = 440$ s, the limiting value of the structural factor is $\varepsilon_{max} = 0.107$. Should the SSTO design be such that $\varepsilon > 0.107$, it would become impossible for the Venture Star spacecraft to reach orbital speed at Space Station altitude. Instead, the spacecraft would reach a suborbital speed at the same altitude.

5. Two-Stage Orbital Spacecraft

The following data were considered for TSTO configurations designed to achieve orbital speed at Space Station altitude, hence $V = 7.633$ km/s at $h = 463$ km:

initial wing loading $W_i / S = 1890$ kgf/m^2, (17a)

initial thrust-to-weight ratio $T_i/W_i = 1.4$. (17b)

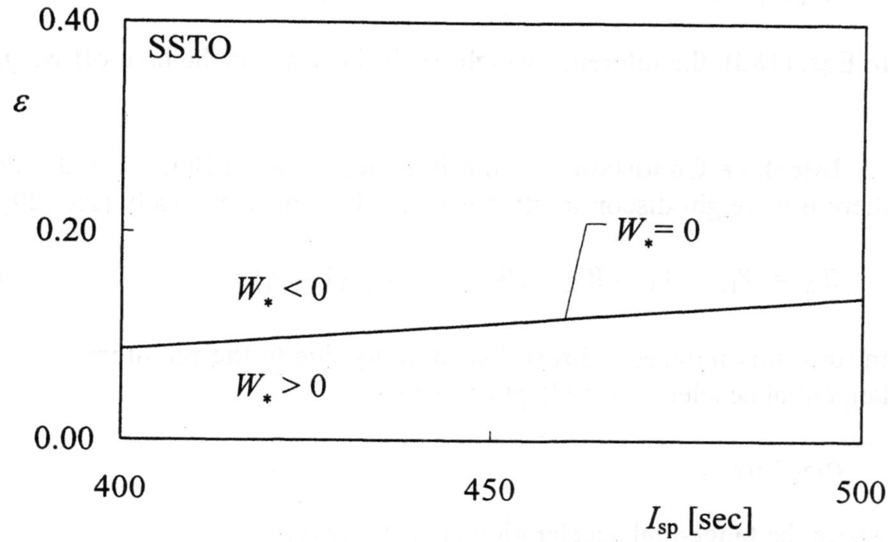

Fig. 2b. SSTO spacecraft, zero-payload line: Structural factor vs specific impulse.

The values (17) are representative of a hypothetical two-stage version of the Venture Star spacecraft.

Let the subscript 1 denote Stage 1; let the subscript 2 denote Stage 2. The maximum payload weight problem was studied first for the case of uniform structural factor, $\varepsilon_1 = \varepsilon_2$, and then for the case of nonuniform structural factor, $\varepsilon_1 < \varepsilon_2$.

5.1. Boundary Conditions. Equations (14), left column, must be understood as initial conditions ($t = 0$, subscript i) for Stage 1; equations (14), right column, must be understood as final conditions ($t = \tau =$ free, subscript f) for Stage 2. Hence,

$$h_{1i} = 0 \text{ km}, \qquad h_{2f} = 463 \text{ km}, \tag{18a}$$

$$V_{1i} = 0 \text{ km/s}, \qquad V_{2f} = 7.633 \text{ km/s}, \tag{18b}$$

$$\gamma_{1i} = \pi/2, \qquad \gamma_{2f} = 0, \tag{18c}$$

$$W_{1i}/W_r = 1, \qquad W_{2f}/W_r = \text{free}. \tag{18d}$$

In Eqs. (18d), the reference weight W_r is the same as the take-off weight.

Interface Conditions. At the interface between Stage 1 and Stage 2, there is a weight discontinuity due to staging, more precisely [see (20)],

$$W_{2i} = W_{1*} = W_{1f} - W_{1s} = (W_{1f} - \varepsilon_1 W_{1i})/(1 - \varepsilon_1)\,. \tag{19a}$$

In turn, this induces a thrust discontinuity due to the requirement that the tangential acceleration be kept unchanged,

$$a_{T2i} = a_{T1f}\,, \tag{19b}$$

where the tangential acceleration is given by (6a).

5.2. Weight Distribution. Relations (10), valid for SSSO and SSTO configurations, are still valid for the TSTO configuration, providing they are rewritten with the subscript 1 for Stage 1 and the subscript 2 for Stage 2.

For Stage 1, the propellant weight, structural weight, and payload weight can be expressed in terms of the initial weight, final weight, and structural factor via the following relations:

$$W_{1p} = W_{1i} - W_{1f}\,, \tag{20a}$$

$$W_{1s} = \varepsilon_1 (W_{1i} - W_{1f}) / (1 - \varepsilon_1)\,, \tag{20b}$$

$$W_{1*} = (W_{1f} - \varepsilon_1 W_{1i}) / (1 - \varepsilon_1)\,, \tag{20c}$$

with

$$\varepsilon_1 = W_{1s} / (W_{1s} + W_{1p})\,, \tag{20d}$$

$$W_{1i} = W_{1p} + W_{1s} + W_{1*}\,. \tag{20e}$$

For Stage 2, the relations analogous to (20) are

$$W_{2p} = W_{2i} - W_{2f} , \tag{21a}$$

$$W_{2s} = \varepsilon_2 (W_{2i} - W_{2f}) / (1 - \varepsilon_2) , \tag{21b}$$

$$W_{2*} = (W_{2f} - \varepsilon_2 W_{2i}) / (1 - \varepsilon_2) , \tag{21c}$$

with

$$\varepsilon_2 = W_{2s} / (W_{2s} + W_{2p}) , \tag{21d}$$

$$W_{2i} = W_{2p} + W_{2s} + W_{2*} . \tag{21e}$$

For the TSTO configuration as a whole, the following relations hold:

$$W_p = W_{1p} + W_{2p} , \tag{22a}$$

$$W_s = W_{1s} + W_{2s} , \tag{22b}$$

$$W_* = W_{2*} , \tag{22c}$$

with

$$W_i = W_{1i} = W_p + W_s + W_* , \tag{22d}$$

$$W_f = W_{2f} . \tag{22e}$$

5.3. Optimization Problem. For the TSTO configuration, the maximum payload weight problem can be formulated as follows [see (21) and (22)]:

$$\max \quad W_* = W_{2*} = W_{2f} - W_{2s} = (W_{2f} - \varepsilon_2 W_{2i}) / (1 - \varepsilon_2), \tag{23a}$$

$$\text{s. t.} \quad (1)\text{-}(6) \text{ and } (18)\text{-}(19). \tag{23b}$$

The unknowns include the state variables h_1, V_1, γ_1, W_1 and h_2, V_2, γ_2, W_2, the control variables α_1, β_1 and α_2, β_2, and the parameters τ_1 and τ_2 which represent the time lengths of Stage 1 and Stage 2. The total time from takeoff to orbit is

$$\tau = \tau_1 + \tau_2. \tag{24}$$

5.4. Computer Runs: Uniform Structural Factor. First, the maximum payload weight problem (23) was solved via SGRA for the following combinations of engine specific impulse and spacecraft structural factor:

$$I_{sp} = 400, 420, 440, 460, 480, 500 \text{ s}, \tag{25a}$$

$$\varepsilon_1 = \varepsilon_2 = \varepsilon = 0.08, 0.10, 0.12. \tag{25b}$$

The results for the normalized final weight W_f/W_i , propellant weight W_p/W_i , structural weight W_s/W_i , and payload weight W_*/W_i associated with various parameter combinations can be found in Refs. 2 and 4. In Fig. 3a, the maximum value of the normalized payload weight is plotted versus the specific impulse for the values (25b) of the structural factor. The main comments are that:

- (i) The normalized payload weight increases as the engine specific impulse increases and as the spacecraft structural factor decreases.
- (ii) The design of TSTO configurations is feasible for each of the parameter combinations considered.
- (iii) For those parameter combinations for which the SSTO configuration is feasible, the TSTO configuration exhibits a much larger payload. As an example, for $I_{sp} = 440$ s and $\varepsilon = 0.10$, the payload of the TSTO configuration (Fig. 3a) is about eight times that of the SSTO configuration (Fig. 2a).

Zero-Payload Line. By proceeding along the lines of Section 3.4, a zero-payload line $W_* = 0$ can be constructed for the TSTO spacecraft with uniform structural factor. With reference to the specific impulse/ structural

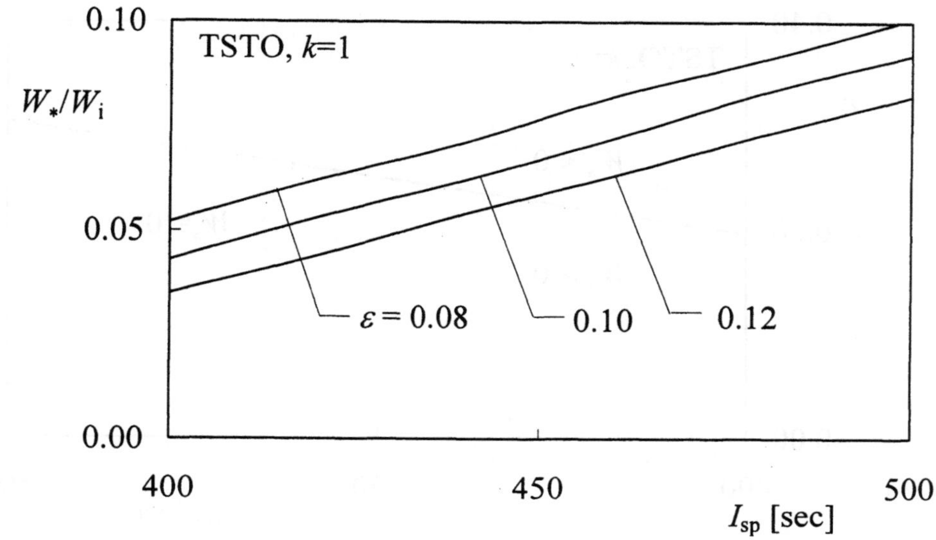

Fig. 3a. TSTO spacecraft, uniform structural factor, k=1: Normalized payload weight vs specific impulse and structural factor.

factor domain, the zero-payload line is shown in Fig. 3b and separates the feasibility region (below) from the unfeasibility region (above). The main comments are that:

(iv) For the TSTO spacecraft, the size of the feasibility region is more than twice that of the SSTO spacecraft.

(v) For a hypothetical two-stage version of the Venture Star spacecraft, with $I_{sp} = 440$ s, the limiting value of the uniform structural factor is $\varepsilon_{max} = 0.239$. This is more than twice the limiting value $\varepsilon_{max} = 0.107$ of the single-stage version of the same spacecraft.

5.5. Computer Runs: Nonuniform Structural Factor. The maximum payload weight problem (23) was solved again via SGRA for the following combinations of engine specific impulse and spacecraft

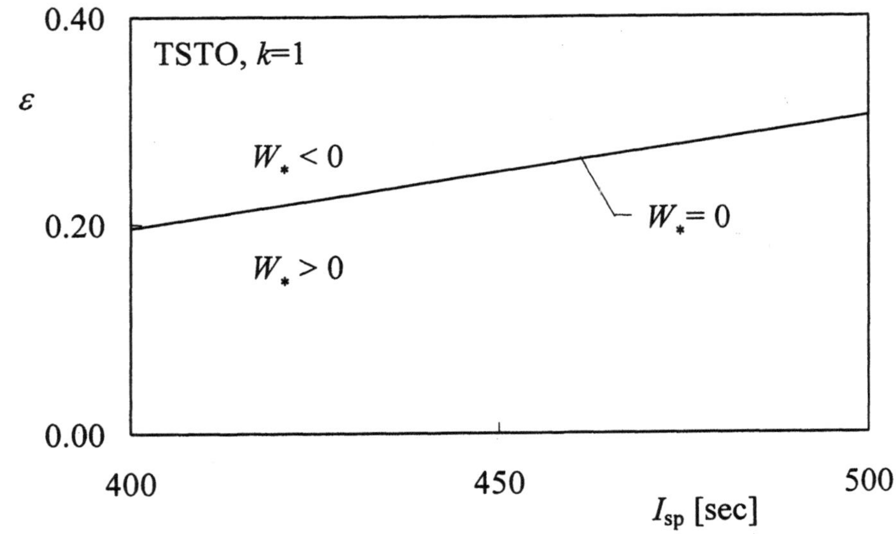

Fig. 3b. TSTO spacecraft, uniform structural factor, k=1, zero-payload line: Structural factor vs specific impulse.

structural factor:

$$I_{sp} = 400, 420, 440, 460, 480, 500 \text{ s}, \tag{26a}$$

$$\varepsilon_1 = \varepsilon, \quad \varepsilon_2 = k\,\varepsilon, \tag{26b}$$

$$\varepsilon = 0.08, 0.10, 0.12, \tag{26c}$$

$$k = 1, 2, 3. \tag{26d}$$

The results for the normalized final weight W_f/W_i , propellant weight W_p/W_i , structural weight W_s/W_i , and payload weight W_*/W_i associated with various parameter combinations can be found in Refs. 3 and 4. In Fig. 4a, the maximum value of the normalized payload weight is plotted versus the specific impulse for the values (26c) of the Stage 1 structural factor and k = 2. In Fig. 4b, the maximum value of the normalized payload weight is plotted versus the specific impulse for $\varepsilon = 0.10$ and the values (26d) of the parameter $k = \varepsilon_2/\varepsilon_1$. The main comments are that:

(i) The normalized payload weight increases as the engine specific impulse increases, as the Stage 1 structural factor decreases, and as the parameter k decreases, hence as the Stage 2 structural factor decreases.

(ii) Even if the Stage 2 structural factor is twice the Stage 1 structural factor ($k = 2$), the TSTO configuration is feasible; this is true for every value of the specific impulse if $\varepsilon = 0.08$ or $\varepsilon = 0.10$ (Fig. 4a) and for $I_{sp} > 412$ s if $\varepsilon = 0.12$.

(iii) For $I_{sp} = 440$ s and $\varepsilon = 0.10$, the maximum value of the parameter k for which feasibility can be guaranteed is $k_{max} = 2.8$ (Fig. 4b); this corresponds to a Stage 2 structural factor $\varepsilon_2 = 0.28$.

Zero-Payload Line. By proceeding along the lines of Section 3.4, zero-payload lines $W_* = 0$ can be constructed for the TSTO spacecraft with nonuniform structural factor. With reference to the specific impulse/structural factor domain, the zero-payload lines are shown in Fig. 4c for the values (26d) of the parameter $k = \varepsilon_2/\varepsilon_1$. For each value of k, these lines separate the feasibility region (below) from the unfeasibility region

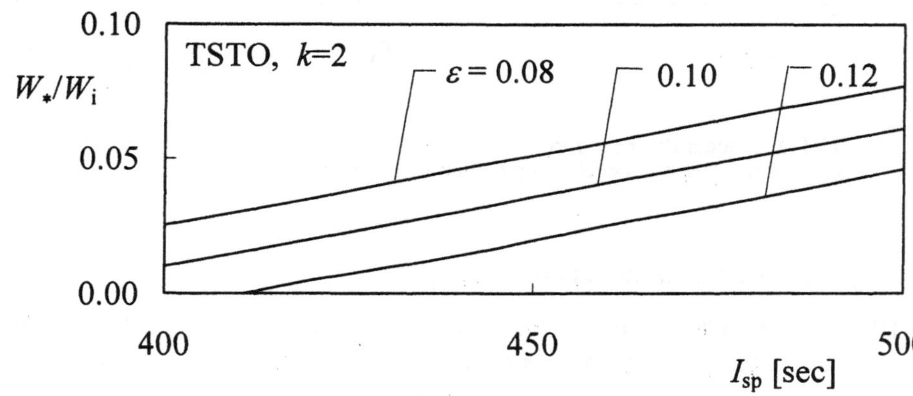

Fig. 4a. TSTO spacecraft, nonuniform structural factor, k=2: Normalized payload weight vs specific impulse and structural factor.

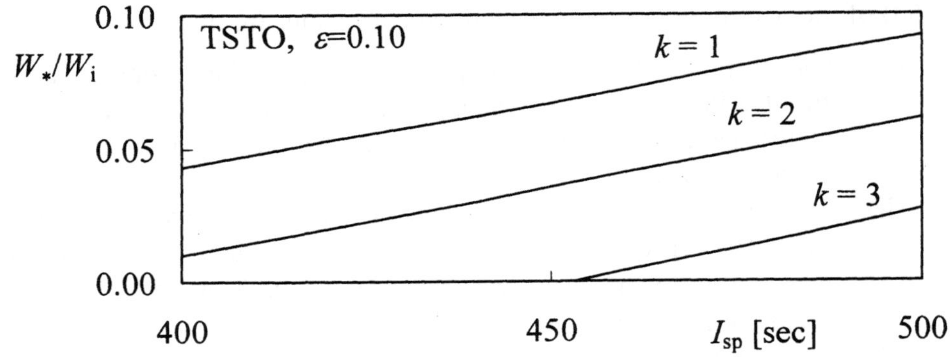

Fig. 4b. TSTO spacecraft, nonuniform structural factor, ε =0.10: Normalized payload weight vs specific impulse and parameter k.

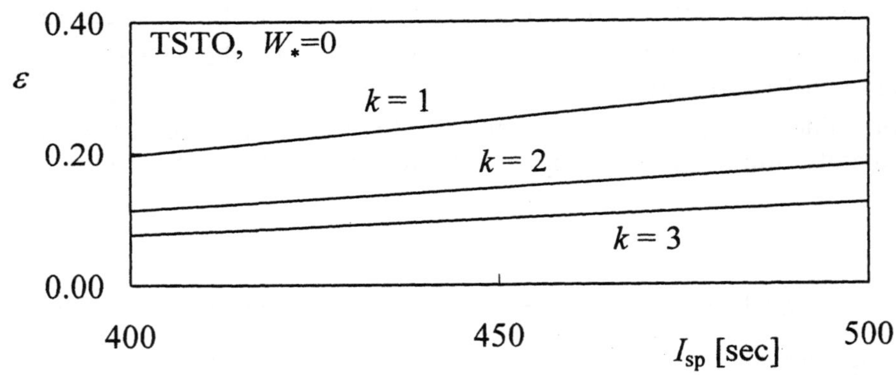

Fig. 4c. TSTO spacecraft, nonuniform structural factor, zero-payload lines: Stage 1 structural factor vs specific impulse and parameter k.

(above). The main comments are that:

(iv) As the parameter k increases, the size of the feasibility region decreases reducing, vis-à-vis the size for $k = 1$, to about 55 percent for k =2 and about 35 percent for $k = 3$.

(v) For $k \cong 2.6$, the zero-payload line of the TSTO spacecraft becomes nearly identical with the zero-payload line of the SSTO spacecraft.

(vi) As a byproduct of (v), let us compare a TSTO configuration (I_{sp}, ε_1, ε_2) with a SSTO configuration (I_{sp}, ε) for the same payload and the same specific impulse. For $\varepsilon_1 = \varepsilon$, one can design a TSTO configuration with ε_2 considerably larger than ε_1, implying increased safety and reliability of the TSTO configuration vis-à-vis the SSTO configuration. The fact that ε_2 can be much larger than ε_1 suggests that an attractive TSTO design might be a first-stage structure made of only tanks and a second-stage structure made of engines, tanks, electronics, and so on.

6. Design Considerations

In Sections 3-5, the maximum payload weight problem was solved for SSSO, SSTO, and TSTO configurations. The results obtained must be taken "cum grano salis" in that they are nonconservative: they disregard the need of propellant for space maneuvers, reentry maneuvers, and reserve margin for emergency. This means that, with reference to the specific impulse/structural factor domain, an actual design must lie wholly inside the feasibility regions of Figs. 1b, 2b, 3b, 4c.

6.1. Structural Factor and Specific Impulse. With the above caveat, the main concept emerging from Sections 3-5 is that the normalized payload weight increases as the engine specific impulse increases and as the spacecraft structural factor decreases. This implies that (i) the use of engines with higher ratio of thrust to propellant weight flow and (ii) the use of lighter materials have a significant effect on payload weight and feasibility of SSSO, SSTO, and TSTO configurations.

6.2. SSSO versus SSTO Configurations. Another concept emerging from Sections 3-4 is that feasibility of the SSSO configuration does not necessarily imply feasibility of the SSTO configuration. The reason for this statement is that the increase in total energy to be imparted to an SSTO configuration is almost 4 times the increase in total energy of an

SSSO configuration performing the task outlined in Section 3. In short, SSSO and SSTO configurations do not belong to the same ballpark; hence, a comparison is not meaningful.

6.3. SSTO versus TSTO Configurations. These configurations do belong to the same ballpark in that they require the same increase in total energy per unit weight to be placed in orbit; hence, a comparison is meaningful.

Figures 5a-5d compare SSTO and TSTO configurations for the case where the latter configuration has uniform structural factor, $\varepsilon_1 = \varepsilon_2 = \varepsilon$. For the Venture Star spacecraft and I_{sp} = 440 s, Fig. 5a shows that, if ε = 0.08, the TSTO payload is about 2.5 times the SSTO payload; Fig. 5b shows that, if ε = 0.10, the TSTO payload is about 8 times the SSTO payload; Fig. 5c shows that, if ε = 0.12, the TSTO spacecraft is feasible with a normalized payload of about 0.05, while the SSTO spacecraft is unfeasible. Figure 5d shows the zero-payload lines of SSTO and TSTO

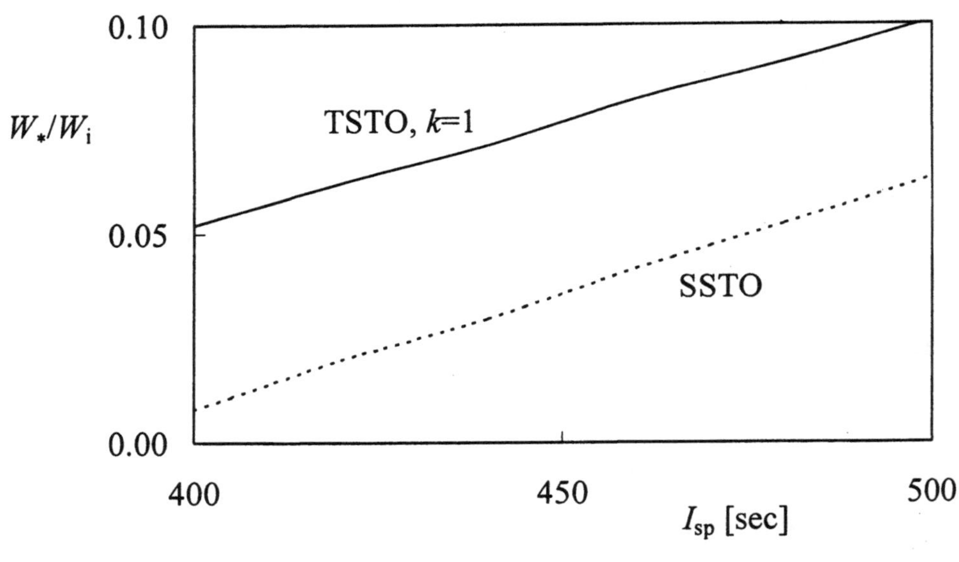

Fig. 5a. SSTO/TSTO comparison, uniform structural factor: Normalized payload weight vs specific impulse for ε=0.08.

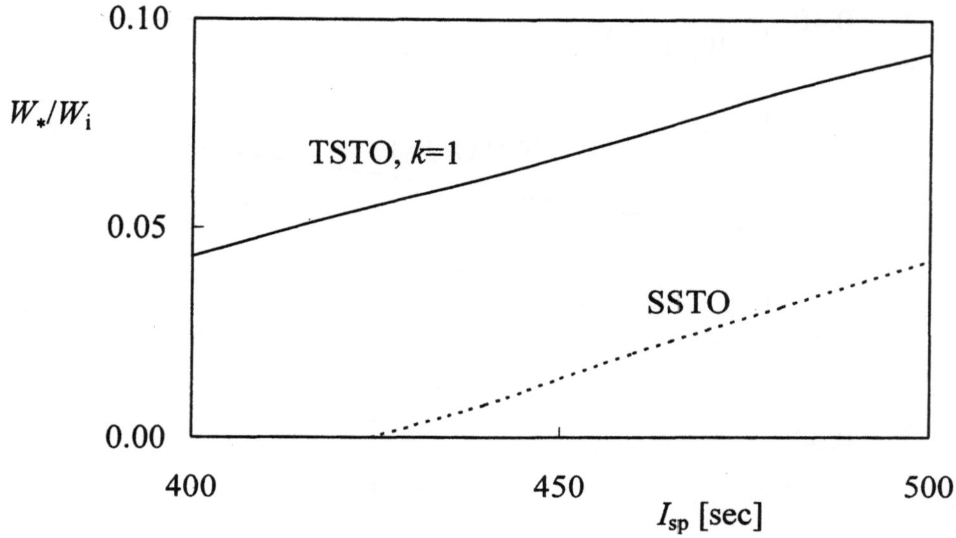

Fig. 5b. SSTO/TSTO comparison, uniform structural factor: Normalized payload weight vs specific impulse for ε=0.10.

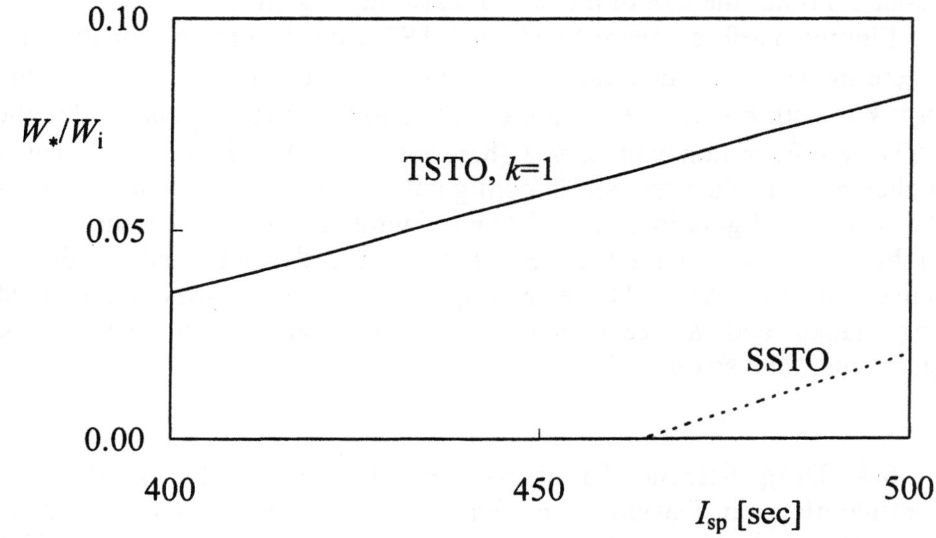

Fig. 5c. SSTO/TSTO comparison, uniform structural factor: Normalized payload weight vs specific impulse for ε=0.12.

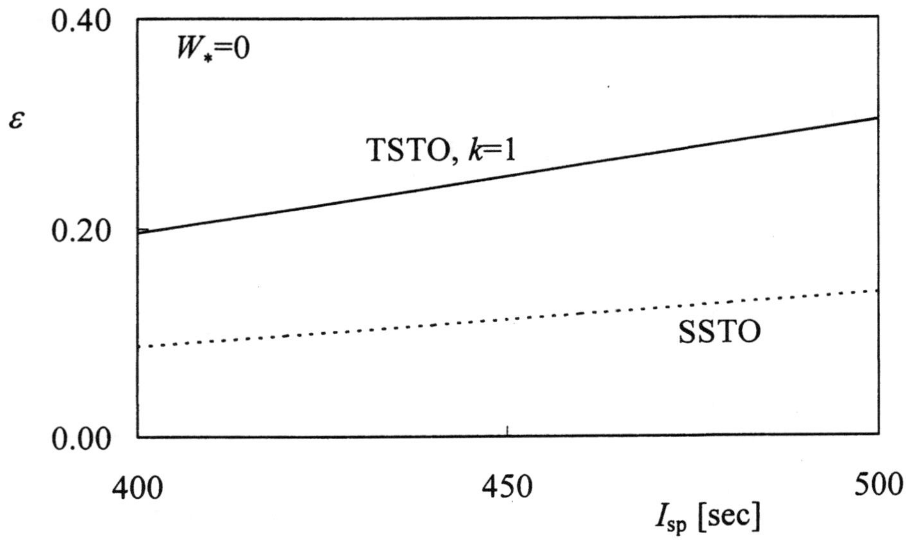

Fig. 5d. SSTO/TSTO comparison, uniform structural factor: Zero-payload lines.

configurations, making clear that the size of the TSTO feasibility region is about 2.5 times the size of the SSTO feasibility region.

Figures 6a-6b compare SSTO and TSTO configurations for the case where the latter configuration has nonuniform structural factor, $\varepsilon_1 = \varepsilon$ and $\varepsilon_2 = k\,\varepsilon$, with $k = 1, 2, 3$. Figure 6a refers to $\varepsilon = 0.10$ and shows that the TSTO configuration with $k = 2$ (hence $\varepsilon_1 = 0.10$ and $\varepsilon_2 = 0.20$) has a higher payload than the SSTO configuration. This implies that, vis-à-vis the SSTO configuration, the TSTO configuration can combine the benefit of higher payload with the benefit of increased safety and reliability. Indeed, an attractive TSTO design might be a first-stage structure made of only tanks and a second-stage structure made of engines, tanks, electronics, and so on.

6.4. Drag Effects. To assess the influence of the aerodynamic configuration on feasibility, a parametric study has been performed. Optimal trajectories have been computed again varying the drag by ± 50%

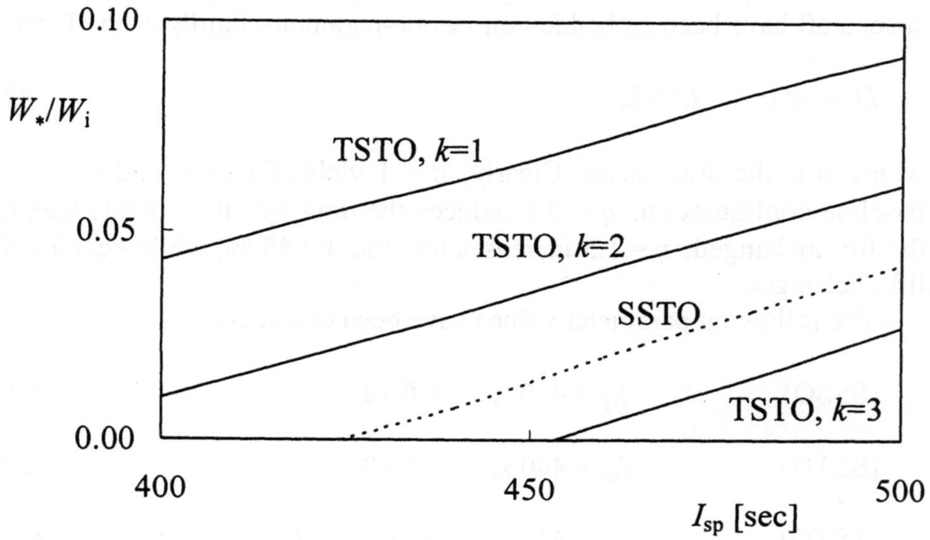

Fig. 6a. SSTO/TSTO comparison, nonuniform structural factor: Normalized payload weight vs specific impulse and parameter k for ε=0.10.

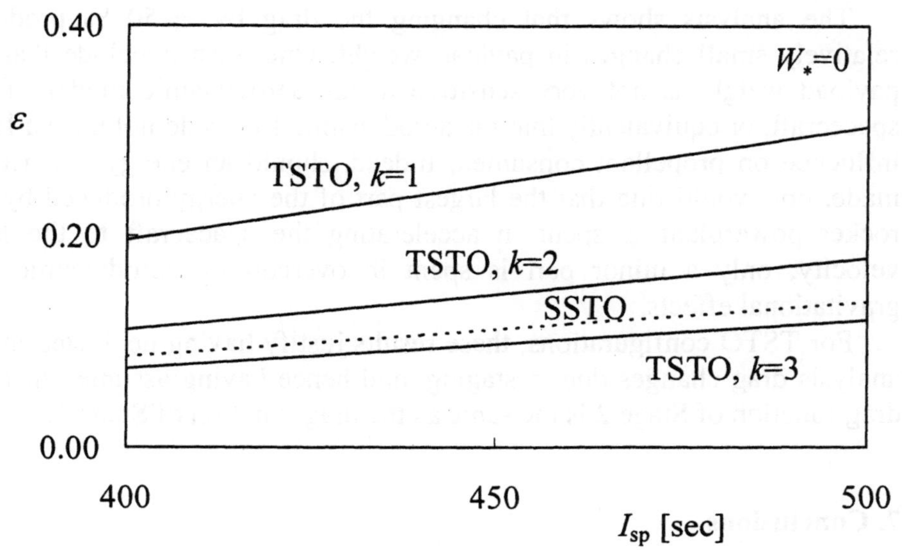

Fig. 6b. SSTO/TSTO comparison, nonuniform structural factor: Zero-payload lines.

while keeping the lift unchanged. Namely, the drag and lift of the spacecraft have been embedded into a one-parameter family of the form

$$D' = \eta D, \qquad L' = L, \tag{27}$$

where η is the drag factor. Clearly, $\eta = 1$ yields the drag and lift of the baseline configuration; $\eta = 0.5$ reduces the drag by 50 %, while keeping the lift unchanged; $\eta = 1.5$ increases the drag by 50 %, while keeping the lift unchanged.

The following parameter values have been considered:

$$\text{(SSSO)} \qquad I_{sp} = 420 \text{ s}, \quad \varepsilon = 0.10, \tag{28a}$$

$$\text{(SSTO)} \qquad I_{sp} = 440 \text{ s}, \quad \varepsilon = 0.10, \tag{28b}$$

$$\text{(TSTO)} \qquad I_{sp} = 440 \text{ s}, \quad \varepsilon_1 = \varepsilon_2 = \varepsilon = 0.10, \tag{28c}$$

with (28c) indicating that a uniform structural factor is being considered for the TSTO configuration. The results are shown in Fig. 7, where the normalized payload weight is plotted versus the drag factor $\eta = D'/D$ for the parameters choices (28).

The analysis shows that changing the drag by ± 50 % produces relatively small changes in payload weight. One must conclude that the payload weight is not very sensitive to the aerodynamic model of the spacecraft, or equivalently that the aerodynamic forces do not have a large influence on propellant consumed. Indeed, should an energy balance be made, one would find that the largest part of the energy produced by the rocket powerplant is spent in accelerating the spacecraft to the final velocity; only a minor part is spent in overcoming aerodynamic and gravitational effects.

For TSTO configurations, these results justify having neglected in the analysis drag changes due to staging, and hence having assumed that the drag function of Stage 2 is the same as the drag function of Stage 1.

7. Conclusions

In this paper, the feasibility of single-stage-suborbital (SSSO), single-

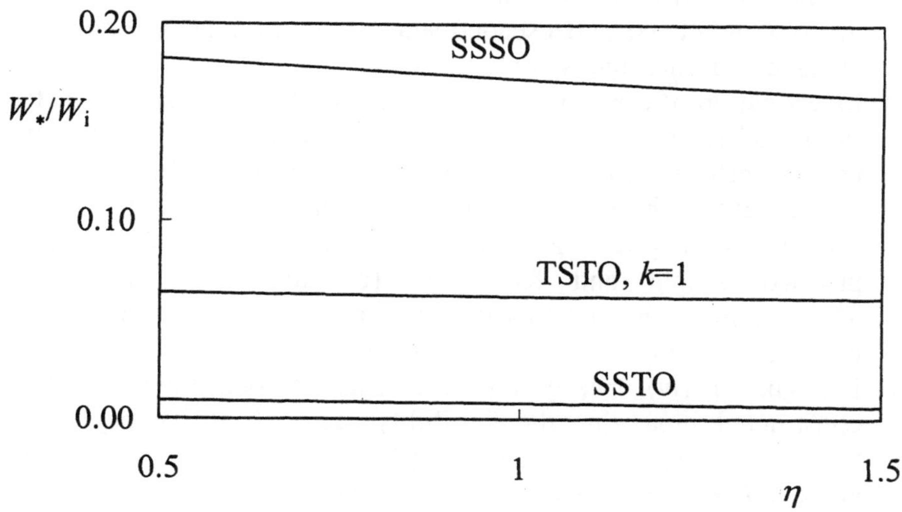

Fig. 7. Normalized payload weight vs drag factor for ε=0.10: SSSO, SSTO, TSTO configurations.

stage-to-orbit (SSTO), and two-stage-to-orbit (TSTO) rocket-powered spacecraft has been investigated using optimal control theory. Ascent trajectories have been optimized for different combinations of spacecraft structural factor and engine specific impulse, the optimization criterion being the maximum payload weight. Normalized payload weights have been computed and used to assess feasibility. The main results are that:

(i) SSSO feasibility does not necessarily imply SSTO feasibility: while SSSO feasibility is guaranteed for all the parameter combinations considered, SSTO feasibility is guaranteed for only certain parameter combinations, which might be beyond the present state of the art.

(ii) For the case of uniform structural factor, not only TSTO feasibility is guaranteed for all the parameter combinations considered, but for the same structural factor a TSTO spacecraft is considerably superior to a SSTO spacecraft in terms of payload weight.

(iii) For the case of nonuniform structural factor, it is possible to design a TSTO spacecraft combining the advantages of higher payload and higher safety/reliability vis-à-vis a SSTO spacecraft.

Indeed, an attractive TSTO design might be a first-stage structure made of only tanks and a second-stage structure made of engines, tanks, electronics, and so on.

(iv) Investigation of areas of potential improvements has shown that: (a) use of lighter materials (smaller spacecraft structural factor) has a significant effect on payload weight and feasibility; (b) use of engines with higher ratio of thrust to propellant weight flow (higher engine specific impulse) has also a significant effect on payload weight and feasibility; (c) on the other hand, aerodynamic improvements via drag reduction have a relatively minor effect on payload weight and feasibility.

(v) In light of (iv), nearly universal zero-payload lines can be constructed separating the feasibility region (positive payload) from the unfeasibility region (negative payload). The zero-payload lines are of considerable help to the designer in assessing the feasibility of a given spacecraft.

(vi) In conclusion, while the design of SSSO spacecraft appears to be feasible, the design of SSTO spacecraft, although attractive from a practical point of view (complete reusability of the spacecraft), might be unfeasible depending on the parameter values considered. Indeed, prudence suggests that TSTO spacecraft be given concurrent consideration, especially if it is not possible to achieve in the near future major improvements in spacecraft structural factor and engine specific impulse.

References

1. MIELE, A., and MANCUSO, S., *Optimal Ascent Trajectories for a Single-Stage Suborbital Spacecraft*, Aero-Astronautics Report 275, Rice University, 1997.

2. MIELE, A., and MANCUSO, S., *Optimal Ascent Trajectories for SSTO and TSTO Spacecraft*, Aero-Astronautics Report 276, Rice University, 1997.

3. MIELE, A., and MANCUSO, S., *Optimal Ascent Trajectories for TSTO Spacecraft: Extensions*, Aero-Astronautics Report 277, Rice University, 1997.

4. MIELE, A., and MANCUSO, S., *Optimal Ascent Trajectories for SSSO, SSTO, and TSTO Spacecraft: Extensions*, Aero-Astronautics Report 278, Rice University, 1997.

5. ANONYMOUS, N. N., *Access to Space Study, Summary Report*, Office of Space Systems Development, NASA Headquarters, 1994.

6. FREEMAN, D. C., TALAY, T. A., STANLEY, D. O., LEPSCH, R. A., and WIHITE, A. W., *Design Options for Advanced Manned Launch Systems*, Journal of Spacecraft and Rockets, Vol.32, No.2, pp.241-249, 1995.

7. GREGORY, I. M., CHOWDHRY, R. S., and McMIMM, J. D., *Hypersonic Vehicle Model and Control Law Development Using H_∞ and μ Synthesis*, Technical Memorandum 4562, NASA, 1994.

8. MIELE, A., WANG, T., and BASAPUR, V.K., *Primal and Dual Formulations of Sequential Gradient-Restoration Algorithms for Trajectory Optimization Problems*, Acta Astronautica, Vol. 13, No. 8, pp. 491-505, 1986.

9. MIELE, A., and WANG, T., *Primal-Dual Properties of Sequential Gradient-Restoration Algorithms for Optimal Control Problems, Part 1: Basic Problem*, Integral Methods in Science and Engineering, Edited by F. R. Payne et al, Hemisphere Publishing Corporation, Washington, DC, pp. 577-607, 1986.

10. MIELE, A., and WANG, T., *Primal-Dual Properties of Sequential Gradient-Restoration Algorithms for Optimal Control Problems, Part 2: General Problem*, Journal of Mathematical Analysis and Applications, Vol. 119, Nos. 1-2, pp. 21-54, 1986.

11. RISHIKOF, B. H., McCORMICK, B. R., PRITCHARD, R. E., and SPONAUGLE, S. J., *SEGRAM: A Practical and Versatile Tool for Spacecraft Trajectory Optimization*, Acta Astronautica, Vol. 26, Nos. 8-10, pp. 599-609, 1992.

12. MIELE, A., and WANG, T., *Optimization and Acceleration Guidance of Flight Trajectories in a Windshear*, Journal of Guidance, Control, and Dynamics, Vol. 10, No. 4, pp.368-377, 1987.

13. MIELE, A., and WANG, T., *Acceleration, Gamma, and Theta*

Guidance for Abort Landing in a Windshear, Journal of Guidance, Control, and Dynamics, Vol. 12, No. 6, pp. 815-821, 1989.

14. MIELE A., LEE, W. Y., and WU, G. D., *Ascent Performance Feasibility of the National Aerospace Plane*, Atti della Accademia delle Scienze di Torino, Vol. 131, pp. 91-108, 1997.

15. MIELE, A., *Recent Advances in the Optimization and Guidance of Aeroassisted Orbital Transfers*, The 1st John V. Breakwell Memorial Lecture, Acta Astronautica, Vol. 38, No. 10, pp. 747-768, 1996.

16. MIELE, A., and WANG, T., *Robust Predictor-Corrector Guidance for Aeroassisted Orbital Transfer*, Journal of Guidance, Control, and Dynamics, Vol. 19, No. 5, pp. 1134-1141, 1996.

17. MIELE, A., *Flight Mechanics, Vol. 1: Theory of Flight Paths*, Chapters 13 and 14, Addison-Wesley Publishing Company, Reading, Massachusetts, 1962.

18. NOAA, NASA, and USAF, *US Standard Atmosphere, 1976*, US Government Printing Office, Washigton, DC, 1976.

2

Design of Moon Missions

A. MIELE[1] AND S. MANCUSO[2]

Abstract. In this paper, a systematic study of the optimization of trajectories for Earth-Moon flight is presented. The optimization criterion is the total characteristic velocity and the parameters to be optimized are: the initial phase angle of the spacecraft with respect to Earth, flight time, and velocity impulses at departure and arrival. The problem is formulated using a simplified version of the restricted three-body model and is solved using the sequential gradient-restoration algorithm for mathematical programming problems.

For given initial conditions, corresponding to a counterclockwise circular low Earth orbit at Space Station altitude, the optimization problem is solved for given final conditions, corresponding to either a clockwise or counterclockwise circular low Moon orbit at different altitudes. Then, the same problem is studied for the Moon-Earth return flight with the same boundary conditions.

The results show that the flight time obtained for the optimal trajectories (about 4.5 days) is larger than that of the Apollo missions (2.5 to 3.2 days). In light of these results, a further parametric study is performed. For given initial and final conditions, the transfer problem is solved again for fixed flight time smaller or larger than the optimal time.

The results show that, if the prescribed flight time is within one

[1] Research Professor and Foyt Professor Emeritus of Engineering, Aerospace Sciences, and Mathematical Sciences, Aero-Astronautics Group, Rice University, Houston, Texas 77005-1892, USA.

[2] Guidance, Navigation, and Control Engineer, European Space Technology and Research Center, 2201 AZ, Nordwijk, Netherlands.

day of the optimal time, the penalty in characteristic velocity is relatively small. For larger time deviations, the penalty in characteristic velocity becomes more severe. In particular, if the flight time is greater than the optimal time by more than two days, no feasible trajectory exists for the given boundary conditions.

The most interesting finding is that the optimal Earth-Moon and Moon-Earth trajectories are mirror images of one another with respect to the Earth-Moon axis. This result extends to optimal trajectories the theorem of image trajectories formulated by Miele for feasible trajectories in 1960.

Key Words. Earth-Moon flight, Moon-Earth flight, Earth-Moon-Earth flight, lunar trajectories, optimal trajectories, astrodyamics, optimization.

1. Introduction

In 1960, the senior author developed the theorem of image trajectories in Earth-Moon space within the frame of the restricted three-body problem (Ref. 1). For both the 2D case and the 3D case, the theorem states that, if a trajectory is feasible in Earth-Moon space, (i) its image with respect to the Earth-Moon axis is also feasible, provided it is flown in the opposite sense. For the 3D case, the theorem guarantees the feasibility of two additional images: (ii) the image with respect to the Moon orbital plane, flown in the same sense as the original trajectory; (iii) the image with respect to the plane containing the Earth-Moon axis and orthogonal to the Moon orbital plane, flown in the opposite sense.

Reference 1 establishes a relation between the outgoing/return trajectories. It is natural to ask whether the feasibility property implies an optimality property. Namely, within the frame of the restricted three-body problem and the 2D case, we inquire whether the image of an optimal Earth-Moon trajectory w.r.t. the Earth-Moon axis has the property of being an optimal Moon-Earth trajectory.

To supply an answer to the above question, we present in this paper a systematic study of optimal Earth-Moon and Moon-Earth trajectories under the following scenario. The optimization criterion is the total characteristic velocity; the class of two-impulse trajectories is considered; the parameters being optimized are four: initial phase angle of spacecraft with respect to either Earth or Moon, flight time, velocity impulse at departure, velocity impulse at arrival.

We study the transfer from a low Earth orbit (LEO) to a low Moon orbit (LMO) and back, with the understanding that the departure from LEO is counterclockwise and the return to LEO is counterclockwise. Concerning LMO, we look at two options: (a) clockwise arrival to LMO, with subsequent clockwise departure from LMO; (b) counterclockwise arrival to LMO, with subsequent counterclockwise departure from LMO. We note that option (a) has characterized all the flights of the Apollo program, and we inquire whether option (b) has any merit.

Finally, because the optimization study reveals that the optimal flight times are considerably larger than the flight times of the Apollo missions, we perform a parametric study by recomputing the LEO-to-LMO and LMO-to-LEO transfers for fixed flight time smaller or larger than the optimal time.

For previous studies related directly or indirectly to the subject under consideration, see Refs. 1-9. References 10-11 are general interest papers. References 12-15 investigate the partial or total use of electric propulsion or nuclear propulsion for Earth-Moon flight. For the algorithms employed to solve the problems formulated in this paper, see Refs. 16-17. For further details on topics covered in this paper, see Ref. 18.

2. System Description

The present study is based on a simplified version of the restricted three-body problem. More precisely, with reference to the motion of a spacecraft in Earth-Moon space, the following assumptions are employed:

(A1) the Earth is fixed in space;
(A2) the eccentricity of the Moon orbit around Earth is neglected;
(A3) the flight of the spacecraft takes place in the Moon orbital plane;
(A4) the spacecraft is subject to only the gravitational fields of Earth and Moon;
(A5) the gravitational fields of Earth and Moon are central and obey the inverse square law;
(A6) the class of two-impulse trajectories, departing with an accelerating velocity impulse tangential to the spacecraft velocity relative to Earth [Moon] and arriving with a braking velocity impulse tangential to the spacecraft velocity relative to Moon [Earth], is considered.

2.1. Differential System. Let the subscripts E, M, P denote the Earth center, Moon center, and spacecraft. Consider an inertial reference frame Exy contained in the Moon orbital plane: its origin is the Earth center; the x-axis points toward the Moon initial position; the y-axis is perpendicular to the x-axis and is directed as the Moon initial inertial velocity. With this understanding, the motion of the spacecraft is described by the following differential system for the position coordinates x_P, y_P and components u_P, w_P of the inertial velocity vector $\vec{V}_P$:

$$\dot{x}_P = u_P \,, \tag{1a}$$

$$\dot{y}_P = w_P \,, \tag{1b}$$

$$\dot{u}_P = -(\mu_E / r_{PE}^3)\, x_P - (\mu_M / r_{PM}^3)\,(x_P - x_M) \,, \tag{1c}$$

$$\dot{w}_P = -(\mu_E / r_{PE}^3)\, y_P - (\mu_M / r_{PM}^3)\,(y_P - y_M) \,, \tag{1d}$$

with

$$V_P = \sqrt{\ } [u_P^2 + w_P^2] \,. \tag{1e}$$

Here μ_E, μ_M are the Earth and Moon gravitational constants; r_{PE}, r_{PM} are the radial distances of the spacecraft from Earth and Moon; x_M, y_M are the Moon inertial coordinates; the dot superscript denotes derivative with respect to the time t, with $0 \leq t \leq \tau$, where 0 is the initial time and τ the final time. The above quantities satisfy the following relations:

$$r_{PE} = \sqrt{\ } [x_P^2 + y_P^2] \,, \tag{2a}$$

$$r_{PM} = \sqrt{\ } [(x_P - x_M)^2 + (y_P - y_M)^2] \,, \tag{2b}$$

$$x_M = r_{ME} \cos\theta_M \,, \tag{2c}$$

$$y_M = r_{ME} \sin\theta_M \,, \tag{2d}$$

$$\theta_M = \omega_M \, t \,. \tag{2e}$$

Here, r_{ME} is the radial distance of the Moon center from the Earth center, θ_M is an angular coordinate associated with the Moon position, more precisely the angle which the vector $\overrightarrow{EM}$ forms with the x-axis; ω_M is the angular velocity of the Moon, assumed constant. Note that, by definition,

$$\theta_M(0) = 0, \quad x_M(0) = r_{ME}, \quad y_M(0) = 0. \tag{3}$$

2.2. Basic Data. The following data are used in the numerical experiments described in this paper:

Earth gravitational constant $\mu_E = 3.986$ E+05 km^3/s^2 , (4a)

Moon gravitational constant $\mu_M = 4.903$ E+03 km^3/s^2 , (4b)

Moon radial distance from Earth center $r_{ME} = 3.844$ E+05 km , (4c)

Moon angular velocity $\omega_M = 2.6491$E–06 rad/s , (4d)

Moon inertial velocity $V_M = 1.0183$ km/s , (4e)

Earth radius $R_E = 6378$ km , (4f)

Moon radius $R_M = 1738$ km . (4g)

2.3. LEO Data. For the low Earth orbit, the following departure data (outgoing trip) and arrival data (return trip) are used in the numerical computation:

$$h_{LEO} = 463 \text{ km} \,, \tag{5a}$$

$$r_{LEO} = 6841 \text{ km} \,, \tag{5b}$$

corresponding to

$$V_{LEO} = 7.633 \text{ km/s} . \tag{5c}$$

The values (5a)-(5b) are the Space Station altitude and corresponding radial distance; the value (5c) is the circular velocity at the Space Station altitude.

2.4. LMO Data. For the low Mars orbit, the following arrival data (outgoing trip) and departure data (return trip) are used in the numerical computation:

$$h_{LMO} = 100, 200, 300 \text{ km} , \tag{6a}$$

$$r_{LMO} = 1838, 1938, 2038 \text{ km} , \tag{6b}$$

corresponding to

$$V_{LMO} = 1.633, 1.590, 1.551 \text{ km/s} . \tag{6c}$$

The values (6a)-(6b) are the LMO altitudes and corresponding radial distances; the values (6c) are the circular velocities at the chosen LMO arrival/departure altitudes.

3. Earth-Moon Flight

We study the LEO-to-LMO transfer of the spacecraft under the following conditions: (i) tangential, accelerating velocity impulse from circular velocity at LEO; (ii) tangential, braking velocity impulse to circular velocity at LMO.

3.1. Departure Conditions. Because of Assumption (A1), Earth fixed in space, the relative-to-Earth coordinates x_{PE}, y_{PE}, u_{PE}, w_{PE} are the same as the inertial coordinates x_P, y_P, u_P, w_P. As a consequence, corresponding to counterclockwise departure from LEO with tangential, accelerating velocity impulse, the departure conditions ($t = 0$) can be written as follows:

$$x_P(0) = x_{PE}(0) = r_{PE}(0)\cos\theta_{PE}(0)\,, \tag{7a}$$

$$y_P(0) = y_{PE}(0) = r_{PE}(0)\sin\theta_{PE}(0)\,, \tag{7b}$$

$$u_P(0) = u_{PE}(0) = -\,V_{PE}(0)\sin\theta_{PE}(0)\,, \tag{7c}$$

$$w_P(0) = w_{PE}(0) = V_{PE}(0)\cos\theta_{PE}(0)\,, \tag{7d}$$

or alternatively,

$$r_{PE}(0) = \sqrt{\{[x_P(0)]^2 + [y_P(0)]^2\}}\,, \tag{8a}$$

$$V_{PE}(0) = \sqrt{\{[u_P(0)]^2 + [w_P(0)]^2\}}\,, \tag{8b}$$

$$x_{PE}(0)\,u_{PE}(0) + y_{PE}(0)\,w_{PE}(0) = 0\,, \tag{8c}$$

where

$$r_{PE}(0) = r_{LEO} = R_E + h_{LEO}\,, \tag{9a}$$

$$V_{PE}(0) = V_{LEO} + \Delta V_{LEO} = \sqrt{[\mu_E / r_{LEO}]} + \Delta V_{LEO}\,. \tag{9b}$$

Here, r_{LEO} is the radius of the low Earth orbit and h_{LEO} is the altitude of the low Earth orbit over the Earth surface; V_{LEO} is the spacecraft velocity in the low Earth orbit (circular velocity) before application of the tangential velocity impulse; ΔV_{LEO} is the accelerating velocity impulse; $V_{PE}(0)$ is the spacecraft velocity after application of the tangential velocity impulse.

Note that Equation (8c) is an orthogonality condition for the vectors $\overrightarrow{EP}(0)$ and $\vec{V}_{PE}(0)$, meaning that the accelerating velocity impulse is tangential to LEO.

3.2. Arrival Conditions. Because Moon is moving with respect to Earth, the relative-to-Moon coordinates x_{PM}, y_{PM}, u_{PM}, w_{PM} are not the same as the inertial coordinates x_P, y_P, u_P, w_P. As a consequence, corresponding to clockwise or counterclockwise arrival to LMO with tangential, braking velocity impulse, the arrival conditions ($t = \tau$) can be written as follows:

$$x_{PM}(\tau) = x_P(\tau) - x_M(\tau) = r_{PM}(\tau) \cos\theta_{PM}(\tau) , \tag{10a}$$

$$y_{PM}(\tau) = y_P(\tau) - y_M(\tau) = r_{PM}(\tau) \sin\theta_{PM}(\tau) , \tag{10b}$$

$$u_{PM}(\tau) = u_P(\tau) - u_M(\tau) = \pm V_{PM}(\tau) \sin\theta_{PM}(\tau) , \tag{10c}$$

$$w_{PM}(\tau) = w_P(\tau) - w_M(\tau) = \mp V_{PM}(\tau) \cos\theta_{PM}(\tau) , \tag{10d}$$

or alternatively,

$$r_{PM}(\tau) = \sqrt{} \{[x_P(\tau) - x_M(\tau)]^2 + [y_P(\tau) - y_M(\tau)]^2\} , \tag{11a}$$

$$V_{PM}(\tau) = \sqrt{} \{[u_P(\tau) - u_M(\tau)]^2 + [w_P(\tau) - w_M(\tau)]^2\} , \tag{11b}$$

$$x_{PM}(\tau)\, u_{PM}(\tau) + y_{PM}(\tau)\, w_{PM}(\tau) = 0 , \tag{11c}$$

where

$$r_{PM}(\tau) = r_{LMO} = R_M + h_{LMO} , \tag{12a}$$

$$V_{PM}(\tau) = V_{LMO} + \Delta V_{LMO} = \sqrt{} [\mu_M / r_{LMO}] + \Delta V_{LMO} , \tag{12b}$$

$$x_M(\tau) = r_{ME} \cos\theta_M(\tau) = r_{ME} \cos(\omega_M \tau) , \tag{12c}$$

$$y_M(\tau) = r_{ME} \sin\theta_M(\tau) = r_{ME} \sin(\omega_M \tau) , \tag{12d}$$

$$u_M(\tau) = - V_M \sin\theta_M(\tau) = - V_M \sin(\omega_M \tau) , \tag{12e}$$

$$w_M(\tau) = V_M \cos\theta_M(\tau) = V_M \cos(\omega_M \tau) . \tag{12f}$$

Here, r_{LMO} is the radius of the low Moon orbit and h_{LMO} is the altitude of the low Moon orbit over the Moon surface; V_{LMO} is the spacecraft velocity in the low Moon orbit (circular velocity) after application of the tangential velocity impulse; ΔV_{LMO} is the braking velocity impulse; $V_{PM}(\tau)$ is the spacecraft velocity before application of the tangential velocity impulse.

In Eqs. (10c)-(10d), the upper sign refers to clockwise arrival to LMO; the lower sign refers to counterclockwise arrival to LMO. Equation (11c) is an orthogonality condition for the vectors $\overrightarrow{MP}(\tau)$ and $\vec{V}_{PM}(\tau)$, meaning that the braking velocity impulse is tangential to LMO.

3.3. Optimization Problem. For Earth-Moon flight, the optimization problem can be formulated as follows: Given the basic data (4) and the terminal data (5)-(6),

$$\min \quad \Delta V = \Delta V_{LEO} + \Delta V_{LMO}, \tag{13a}$$

$$\text{s. t.} \quad (1)\text{-}(2) \text{ and } (7)\text{-}(12), \tag{13b}$$

where ΔV is the total characteristic velocity. The unknowns include the state variables $x_P(t)$, $y_P(t)$, $u_P(t)$, $w_P(t)$ and the parameters ΔV_{LEO}, ΔV_{LMO}, $\theta_{PE}(0)$, τ.

While this problem can be treated as either a mathematical programming problem or an optimal control problem, the former point of view is employed here because of its simplicity. In the mathematical programming formulation, the main function of the differential system (1)-(2) is that of connecting the initial point with the final point and in particular supplying the gradients of the final conditions with respect to the initial conditions and/or problem parameters. In the particular case, because the problem parameters determine completely the initial conditions, the gradients are formed only with respect to the problem parameters.

To sum up, we have a mathematical programming problem in which the minimization of the performance index (13a) is sought with respect to the values of ΔV_{LEO}, ΔV_{LMO}, $\theta_{PE}(0)$, τ which satisfy the radius condition (11a)-(12a), circularization condition (11b)-(12b), and tangency condition (10)-(11c). Since we have $n = 4$ parameters and $q = 3$ constraints, the number of degrees of freedom is $n - q = 1$. Therefore, it is appropriate to employ the sequential gradient-restoration algorithm (SGRA) for mathematical programming problems (Ref. 16).

3.4. Results. Two groups of optimal trajectories have been computed. The first group is formed by trajectories for which the arrival to LMO is clockwise; the second group is formed by trajectories for which the arrival to LMO is counterclockwise. For h_{LMO} = 100 km, the results are shown in Tables 1-2 and Figs. 1-2. The major parameters of the problem, the phase angles at departure, and the phase angles at arrival are shown in Table 1 for clockwise LMO arrival and Table 2 for counterclockwise LMO arrival.

Table 1a. Earth-Moon flight, h_{LMO} = 100 km, clockwise LMO arrival, major parameters.

ΔV [km/s]	ΔV_{LEO} [km/s]	ΔV_{LMO} [km/s]	$\theta_{PE}(0)$ [deg]	τ [days]
3.882	3.068	0.814	−116.88	4.50

Table 1b. Earth-Moon flight, h_{LMO} = 100 km, clockwise LMO arrival, phase angles at departures.

$\theta_{PE}(0)$ [deg]	$\theta_M(0)$ [deg]	$\phi_{PE}(0)$ [deg]	Note
−116.88	0.00	−116.88	See Eqs. (21)

Table 1c. Earth-Moon flight, h_{LMO} = 100 km, clockwise LMO arrival, phase angles at arrival.

$\theta_{PM}(\tau)$ [deg]	$\theta_M(\tau)$ [deg]	$\phi_{PM}(\tau)$ [deg]	Note
14.37	59.04	−44.67	See Eqs. (21)

Table 2a. Earth-Moon flight, h_{LMO} = 100 km, counterclockwise LMO arrival, major parameters.

ΔV [km/s]	ΔV_{LEO} [km/s]	ΔV_{LMO} [km/s]	$\theta_{PE}(0)$ [deg]	τ [days]
3.876	3.065	0.811	−118.98	4.37

Table 2b. Earth-Moon flight, h_{LMO} = 100 km, counterclockwise LMO arrival, phase angles at departure.

$\theta_{PE}(0)$ [deg]	$\theta_{M}(0)$ [deg]	$\phi_{PE}(0)$ [deg]	Note
−118.98	0.00	−118.98	See Eqs. (21)

Table 2c. Earth-Moon flight, h_{LMO} = 100 km, counterclockwise LMO arrival, phase angles at arrival.

$\theta_{PM}(\tau)$ [deg]	$\theta_{M}(\tau)$ [deg]	$\phi_{PM}(\tau)$ [deg]	Note
−60.87	57.28	−118.15	See Eqs. (21)

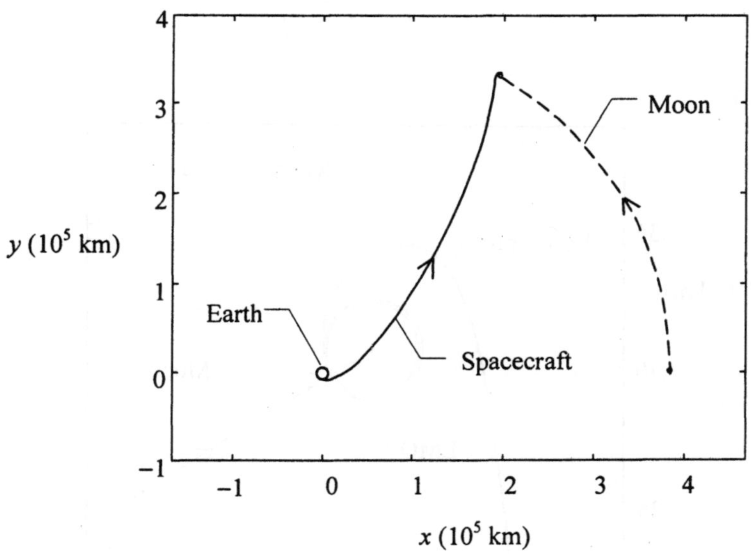

Fig. 1a. Earth-Moon trajectory, inertial coordinate system, h_{LMO} = 100 km, clockwise LMO arrival, Earth-Moon space.

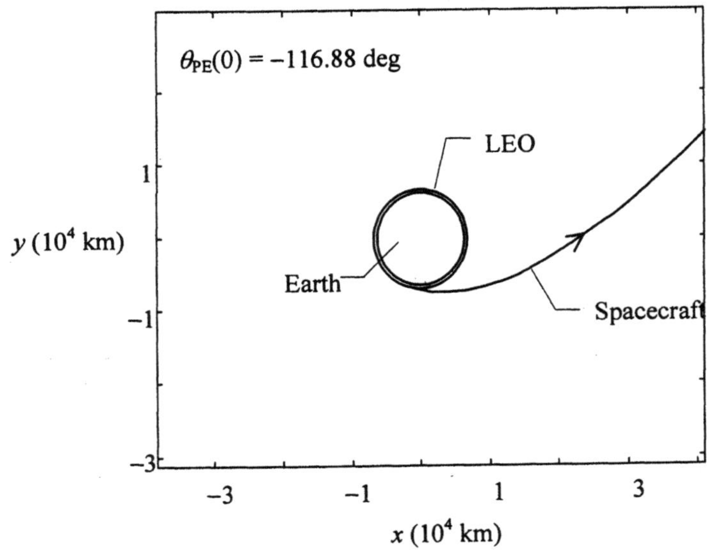

Fig. 1b. Earth-Moon trajectory, inertial coordinate system, $h_{LMO} = 100$ km, clockwise LM^ arrival, near-Earth space.

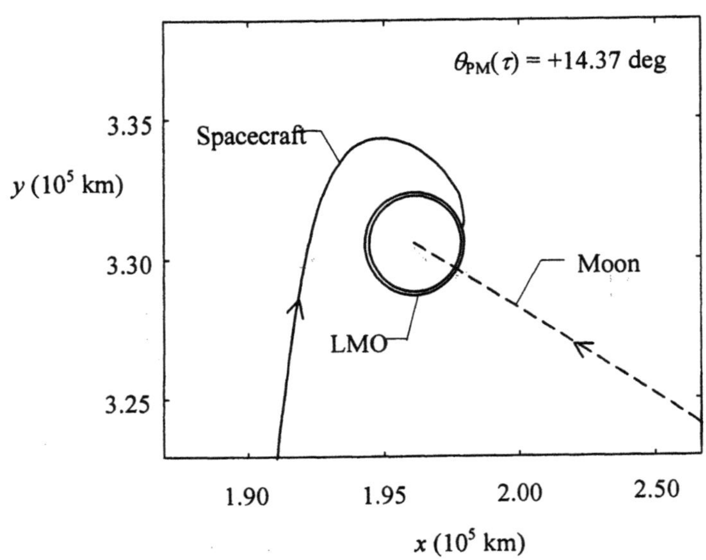

Fig. 1c. Earth-Moon trajectory, inertial coordinate system, $h_{LMO} = 100$ km, clockwise LMO arrival, near-Moon space.

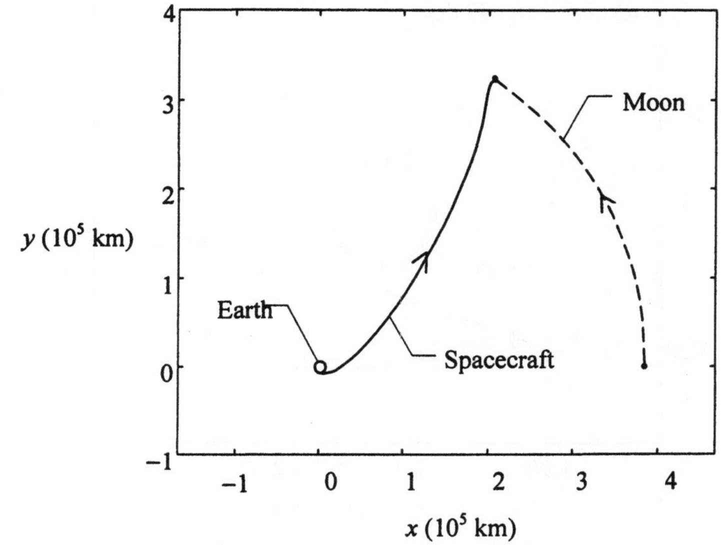

Fig. 2a. Earth-Moon trajectory, inertial coordinate system, $h_{LMO} = 100$ km, counterclockwise LMO arrival, Earth-Moon space.

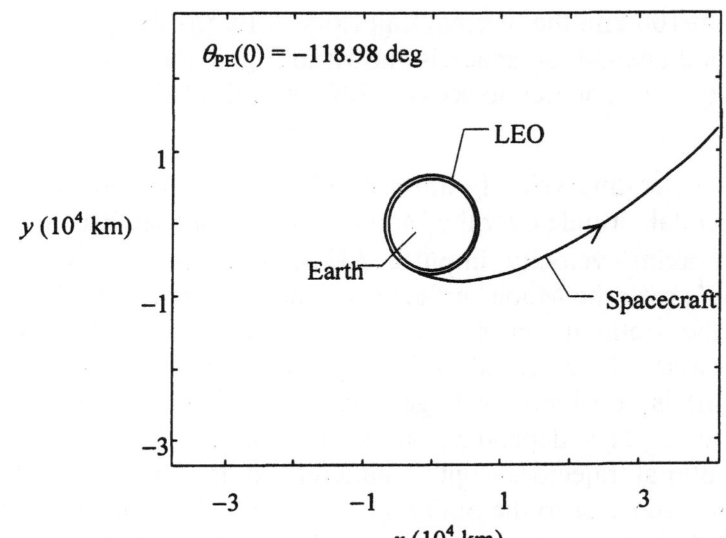

Fig. 2b. Earth-Moon trajectory, inertial coordinate system, $h_{LMO} = 100$ km, counterclockwise LMO arrival, near-Earth space.

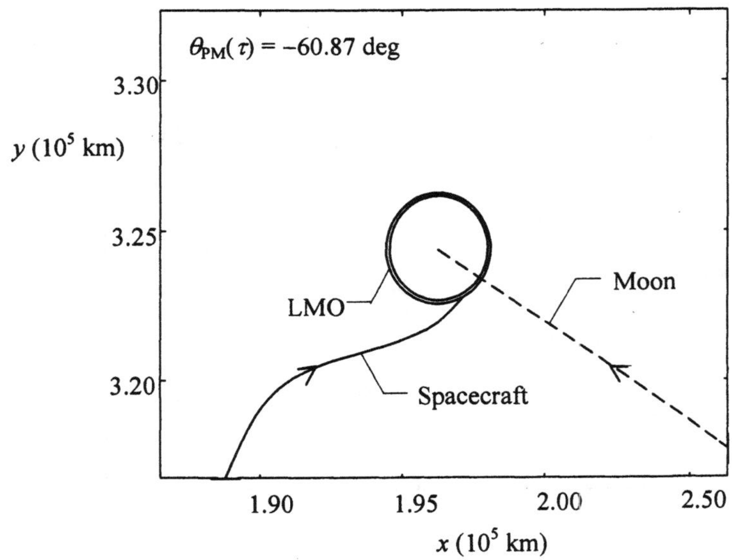

Fig. 2c. Earth-Moon trajectory, inertial coordinate system, $h_{LMO} = 100$ km, counterclockwise LMO arrival, near-Moon space.

Also for $h_{LMO} = 100$ km, the optimal trajectory in Earth-Moon space, near-Earth space, and near-Moon space is shown in Fig. 1 for clockwise LMO arrival and Fig. 2 for counterclockwise LMO arrival. Major comments are as follows:

(i) the accelerating velocity impulse ΔV_{LEO} is nearly independent of the orbital altitude over the Moon surface h_{LMO} (see Ref. 18);

(ii) the braking velocity impulse ΔV_{LMO} decreases as the orbital altitude over the Moon surface h_{LMO} increases (see Ref. 18);

(iii) for the optimal trajectories, the flight time (4.50 days for clockwise LMO arrival, 4.37 days for counterclockwise LMO arrival) is considerably larger than that of the Apollo missions (2.5 to 3.2 days, depending on the mission);

(iv) the optimal trajectories with counterclockwise arrival to LMO are slightly superior to the optimal trajectories with clockwise arrival to LMO in terms of characteristic velocity and flight time.

4. Moon-Earth Flight

We study the LMO-to-LEO transfer of the spacecraft under the following conditions: (i) tangential, accelerating velocity impulse from circular velocity at LMO; (ii) tangential, braking velocity impulse to circular velocity at LEO.

4.1. Departure Conditions. Because Moon is moving with respect to Earth, the relative-to-Moon coordinates x_{PM}, y_{PM}, u_{PM}, w_{PM} are not the same as the inertial coordinates x_P, y_P, u_P, w_P. As a consequence, corresponding to clockwise or counterclockwise departure from LMO with tangential, accelerating velocity impulse, the departure conditions (t = 0) can be written as follows:

$$x_{PM}(0) = x_P(0) - x_M(0) = r_{PM}(0)\cos\theta_{PM}(0)\,, \tag{14a}$$

$$y_{PM}(0) = y_P(0) - y_M(0) = r_{PM}(0)\sin\theta_{PM}(0)\,, \tag{14b}$$

$$u_{PM}(0) = u_P(0) - u_M(0) = \pm\, V_{PM}(0)\sin\theta_{PM}(0)\,, \tag{14c}$$

$$w_{PM}(0) = w_P(0) - w_M(0) = \mp\, V_{PM}(0)\cos\theta_{PM}(0)\,, \tag{14d}$$

or alternatively,

$$r_{PM}(0) = \sqrt{\{[x_P(0) - x_M(0)]^2 + [y_P(0) - y_M(0)]^2\}}\,, \tag{15a}$$

$$V_{PM}(0) = \sqrt{\{[u_P(0) - u_M(0)]^2 + [w_P(0) - w_M(0)]^2\}}\,, \tag{15b}$$

$$x_{PM}(0)\, u_{PM}(0) + y_{PM}(0)\, w_{PM}(0) = 0\,, \tag{15c}$$

where

$$r_{PM}(0) = r_{LMO} = R_M + h_{LMO}\,, \tag{16a}$$

$$V_{PM}(0) = V_{LMO} + \Delta V_{LMO} = \sqrt{[\mu_M / r_{LMO}]} + \Delta V_{LMO}\,, \tag{16b}$$

$$x_M(0) = r_{ME} \cos\theta_M(0) = r_{ME} \cos(\omega_M\, 0) = r_{ME}\,, \tag{16c}$$

$$y_M(0) = r_{ME} \sin\theta_M(0) = r_{ME} \sin(\omega_M\, 0) = 0\,, \tag{16d}$$

$$u_M(0) = -\, V_M \sin\theta_M(0) = -\, V_M \sin(\omega_M\, 0) = 0\,, \tag{16e}$$

$$w_M(0) = V_M \cos\theta_M(0) = V_M \cos(\omega_M\, 0) = V_M\,. \tag{16f}$$

Here, r_{LMO} is the radius of the low Moon orbit and h_{LMO} is the altitude of the low Moon orbit over the Moon surface; V_{LMO} is the spacecraft velocity in the low Moon orbit (circular velocity) before application of the tangential velocity impulse; ΔV_{LMO} is the accelerating velocity impulse; $V_{PM}(0)$ is the spacecraft velocity after application of the tangential velocity impulse.

In Eqs. (14c)-(14d), the upper sign refers to clockwise departure from LMO; the lower sign refers to counterclockwise departure from LMO. Equation (15c) is an orthogonality condition for the vectors $\overrightarrow{MP}(0)$ and $\vec{V}_{PM}(0)$, meaning that the accelerating velocity impulse is tangential to LMO.

4.2. Arrival Conditions. Because of Assumption (A1), Earth fixed in space, the relative-to-Earth coordinates x_{PE}, y_{PE}, u_{PE}, w_{PE} are the same as the inertial coordinates x_P, y_P, u_P, w_P. As a consequence, corresponding to counterclockwise arrival to LEO with tangential, braking velocity impulse, the arrival conditions ($t = \tau$) can be written as follows:

$$x_P(\tau) = x_{PE}(\tau) = r_{PE}(\tau) \cos\theta_{PE}(\tau)\,, \tag{17a}$$

$$y_P(\tau) = y_{PE}(\tau) = r_{PE}(\tau) \sin\theta_{PE}(\tau)\,, \tag{17b}$$

$$u_P(\tau) = u_{PE}(\tau) = -\, V_{PE}(\tau) \sin\theta_{PE}(\tau)\,, \tag{17c}$$

$$w_P(\tau) = w_{PE}(\tau) = V_{PE}(\tau) \cos\theta_{PE}(\tau)\,, \tag{17d}$$

or alternatively,

$$r_{PE}(\tau) = \sqrt{\{[x_P(\tau)]^2 + [y_P(\tau)]^2\}}\,, \tag{18a}$$

$$V_{PE}(\tau) = \sqrt{\{[u_P(\tau)]^2 + [w_P(\tau)]^2\}}\,, \tag{18b}$$

$$x_{PE}(\tau)\, u_{PE}(\tau) + y_{PE}(\tau)\, w_{PE}(\tau) = 0\,, \tag{18c}$$

where

$$r_{PE}(\tau) = r_{LEO} = R_E + h_{LEO}\,, \tag{19a}$$

$$V_{PE}(\tau) = V_{LEO} + \Delta V_{LEO} = \sqrt{[\mu_E / r_{LEO}]} + \Delta V_{LEO}\,. \tag{19b}$$

Here, r_{LEO} is the radius of the low Earth orbit and h_{LEO} is the altitude of the low Earth orbit over the Earth surface; V_{LEO} is the spacecraft velocity in the low Earth orbit (circular velocity) after application of the tangential velocity impulse; ΔV_{LEO} is the braking velocity impulse; $V_{PE}(\tau)$ is the spacecraft velocity before application of the tangential velocity impulse.

Note that Equation (18c) is an orthogonality condition for the vectors $\overrightarrow{EP}(\tau)$ and $\vec{V}_{PE}(\tau)$, meaning that the braking velocity impulse is tangential to LEO.

4.3. Optimization Problem. For Moon-Earth flight, the optimization problem can be formulated as follows: Given the basic data (4) and the terminal data (5)-(6),

$$\min \quad \Delta V = \Delta V_{LMO} + \Delta V_{LEO}\,, \tag{20a}$$

$$\text{s. t.} \quad (1)\text{-}(2) \text{ and } (14)\text{-}(19)\,, \tag{20b}$$

where ΔV is the total characteristic velocity. The unknowns include the state variables $x_P(t)$, $y_P(t)$, $u_P(t)$, $w_P(t)$ and the parameters ΔV_{LMO}, ΔV_{LEO}, $\theta_{PM}(0)$, τ.

Similarly to what is stated in Section 3.3, we are in the presence of a mathematical programming problem in which the minimization of the performance index (20a) is sought with respect to the values of ΔV_{LMO}, ΔV_{LEO}, $\theta_{PM}(0)$, τ which satisfy the radius condition (18a)-(19a), circularization condition (18b)-(19b), and tangency condition (17)-(18c). Once more, we have $n = 4$ parameters and $q = 3$ constraints, so that the number of degrees of freedom is $n - q = 1$. Therefore, it is appropriate to employ the sequential gradient-restoration algorithm (SGRA) for mathematical programming problems (Ref. 16).

4.4. Results. Two groups of optimal trajectories have been computed. The first group is formed by trajectories for which the departure from LMO is clockwise; the second group is formed by trajectories for which the departure from LMO is counterclockwise. The results are presented in Tables 3-4 and Figs. 3-4. For h_{LMO} = 100 km, the major parameters of the problem, the phase angles at departure, and the phase angles at arrival are shown in Table 3 for clockwise LMO departure and Table 4 for counterclockwise LMO departure. Also for h_{LMO} = 100 km, the optimal

Table 3a. Moon-Earth flight, h_{LMO} = 100 km, clockwise LMO departure, major parameters.

ΔV [km/s]	ΔV_{LEO} [km/s]	ΔV_{LMO} [km/s]	$\theta_{PM}(0)$ [deg]	τ [days]
3.882	3.068	0.814	44.67	4.50

Table 3b. Moon-Earth flight, h_{LMO} = 100 km, clockwise LMO departure, phase angles at departures.

$\theta_{PM}(0)$ [deg]	$\theta_M(0)$ [deg]	$\phi_{PM}(0)$ [deg]	Note
44.67	0.00	44.67	See Eqs. (21)

Table 3c. Moon-Earth flight, h_{LMO} = 100 km, clockwise LMO departure, phase angles at arrival.

$\theta_{PE}(\tau)$ [deg]	$\theta_M(\tau)$ [deg]	$\phi_{PE}(\tau)$ [deg]	Note
175.92	59.04	116.88	See Eqs. (21)

Table 4a. Moon-Earth flight, h_{LMO} = 100 km, counterclockwise LMO departure, major parameters.

ΔV [km/s]	ΔV_{LEO} [km/s]	ΔV_{LMO} [km/s]	$\theta_{PM}(0)$ [deg]	τ [days]
3.876	3.065	0.811	118.15	4.37

Table 4b. Moon-Earth flight, h_{LMO} = 100 km, counterclockwise LMO departure, phase angles at departure.

$\theta_{PM}(0)$ [deg]	$\theta_M(0)$ [deg]	$\phi_{PM}(0)$ [deg]	Note
118.15	0.00	118.15	See Eqs. (21)

Table 4c. Moon-Earth flight, h_{LMO} = 100 km, counterclockwise LMO departure, phase angles at arrival.

$\theta_{PE}(\tau)$ [deg]	$\theta_M(\tau)$ [deg]	$\phi_{PE}(\tau)$ [deg]	Note
176.26	57.28	118.98	See Eqs. (21)

trajectory in Moon-Earth space, near-Moon space, and near-Earth space is shown in Fig. 3 for clockwise LMO departure and Fig. 4 for counterclockwise LMO departure. Major comments are as follows:

(i) the accelerating velocity impulse ΔV_{LMO} decreases as the orbital altitude over the Moon surface h_{LMO} increases (see Ref. 18);

(ii) the braking velocity impulse ΔV_{LEO} is nearly independent of the orbital altitude over the Moon surface h_{LMO} (see Ref. 18);

(iii) for the optimal trajectories, the flight time (4.50 days for clockwise LMO departure, 4.37 days for counterclockwise LMO departure) is considerably larger than that of the Apollo missions (2.5 to 3.2 days, depending on the mission);

(iv) the optimal trajectories with counterclockwise departure from LMO are slightly superior to the optimal trajectories with clockwise departure from LMO in terms of characteristic velocity and flight time.

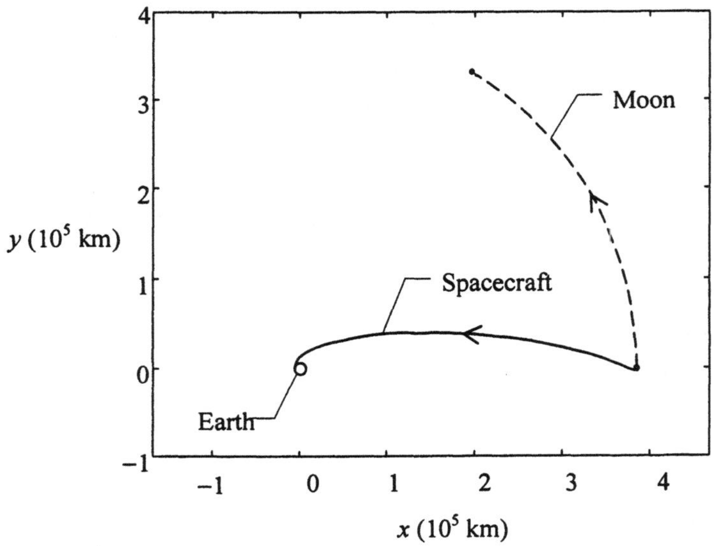

Fig.3a. Moon-Earth trajectory, inertial coordinate system, h_{LMO} = 100 km, clockwise LMO departure, Earth-Moon space.

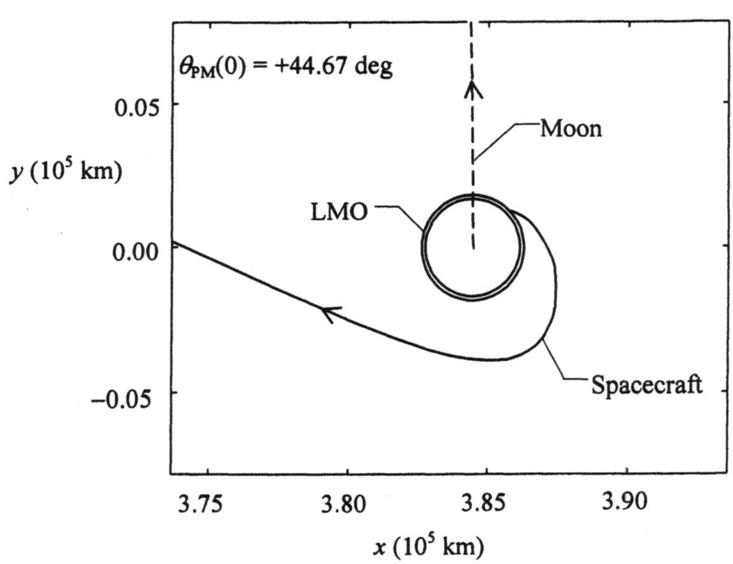

Fig. 3b. Moon-Earth trajectory, inertial coordinate system, h_{LMO} = 100 km, clockwise LMO departure, near-Moon space.

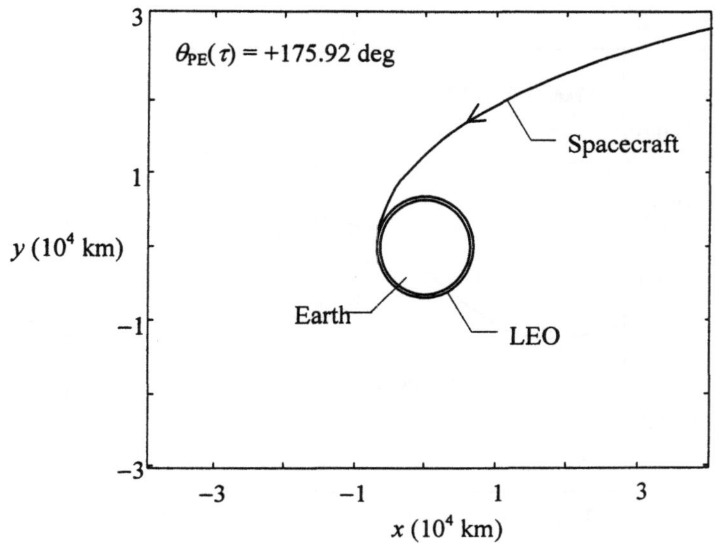

Fig. 3c. Moon-Earth trajectory, inertial coordinate system, h_{LMO} = 100 km, clockwise LMO departure, near-Earth space.

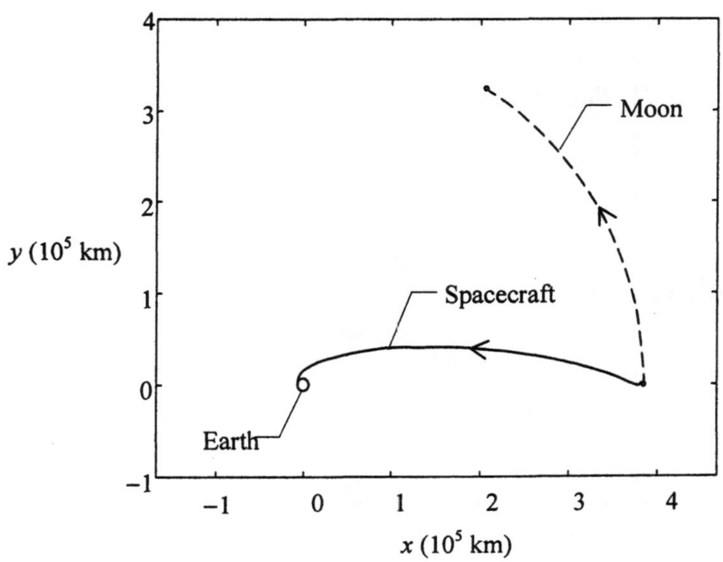

Fig. 4a. Moon-Earth trajectory, inertial coordinate system, h_{LMO} = 100 km, counterclockwise LMO departure, Earth-Moon space.

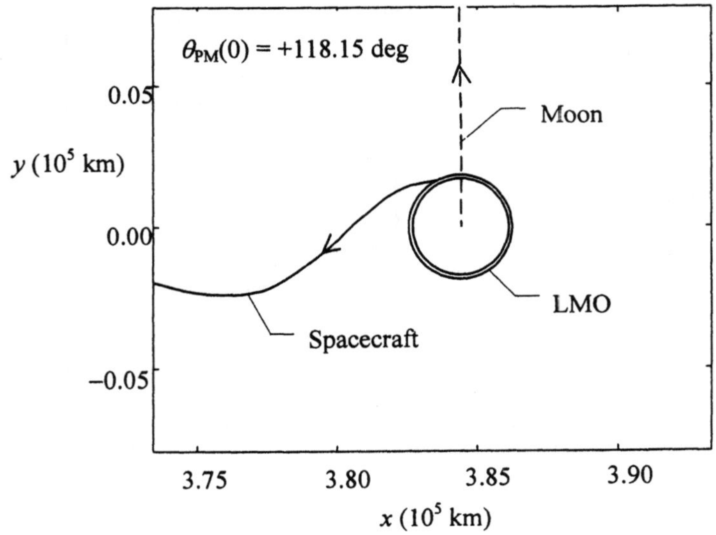

Fig. 4b. Moon-Earth trajectory, inertial coordinate system, $h_{LMO} = 100$ km, counterclockwise LMO departure, near-Moon space.

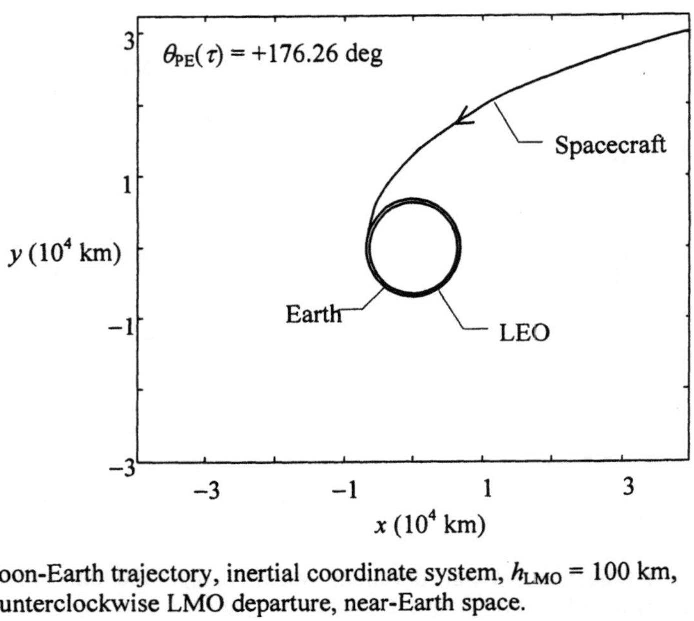

Fig. 4c. Moon-Earth trajectory, inertial coordinate system, $h_{LMO} = 100$ km, counterclockwise LMO departure, near-Earth space.

5. Earth-Moon-Earth Flight

A very interesting observation can be made by comparing the results obtained in Sections 3 and 4, in particular Tables 1-2 and Tables 3-4. In these tables, two kinds of phase angles are reported: for the phase angles θ_{PE} and θ_{PM}, the reference line is the initial direction of the Earth-Moon axis; for the phase angles ϕ_{PE} and ϕ_{PM}, the reference line is the instantaneous direction of the Earth-Moon axis. The relations leading from the θ angles to the ϕ angles are given below,

$$\phi_{PE}(t) = \theta_{PE}(t) - \theta_{M}(t) , \tag{21a}$$

$$\phi_{PM}(t) = \theta_{PM}(t) - \theta_{M}(t) . \tag{21b}$$

Thus, $\phi_{PE}(t)$ is the angle which the vector $\overrightarrow{EP}$ forms with the rotating Earth-Moon axis, while $\phi_{PM}(t)$ is the angle which the vector $\overrightarrow{MP}$ forms with the rotating Earth-Moon axis.

With the above definitions in mind, let the departure point of the outgoing trip be paired with the arrival point of the return trip; conversely, let the departure point of the return trip be paired with the arrival point of the outgoing trip. For these paired points, the following relations hold (see Tables 1-4):

$$[\phi_{PE}(0)]_{outgoing} + [\phi_{PE}(\tau)]_{return} = 0 , \tag{22a}$$

$$[\phi_{PM}(0)]_{return} + [\phi_{PM}(\tau)]_{outgoing} = 0 , \tag{22b}$$

showing that, for the optimal outgoing/return trajectories and in a rotating coordinate system, corresponding phase angles are equal in modulus and opposite in sign, consistently with the predictions of the theorem of the image trajectories formulated by Miele for feasible trajectories in 1960 (Ref. 1).

To better visualize this result, the optimal trajectories of Sections 3 and 4, which were plotted in Figs. 1-4 in an inertial coordinate system Exy, have been replotted in Figs. 5-6 in a rotating coordinate system E$\xi\eta$; here, the origin is the Earth center, the ξ-axis coincides with the instantaneous Earth-Moon axis and is directed from Earth to Moon; the η-axis is perpendicular to the ξ-axis and is directed as the Moon inertial velocity.

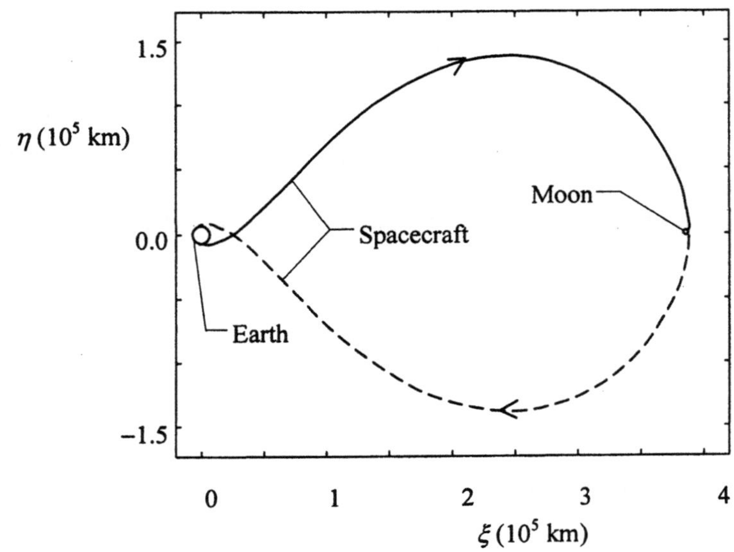

Fig. 5a. Earth-Moon-Earth trajectory, rotating coordinate system, h_{LMO} = 100 km, clockwise LMO, Earth-Moon space.

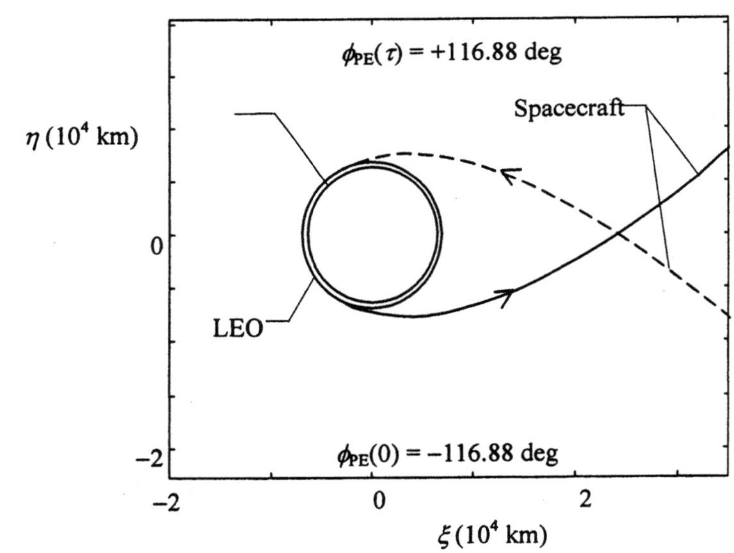

Fig. 5b. Earth-Moon-Earth trajectory, rotating coordinate system, h_{LMO} = 100 km, clockwise LMO, near-Earth space.

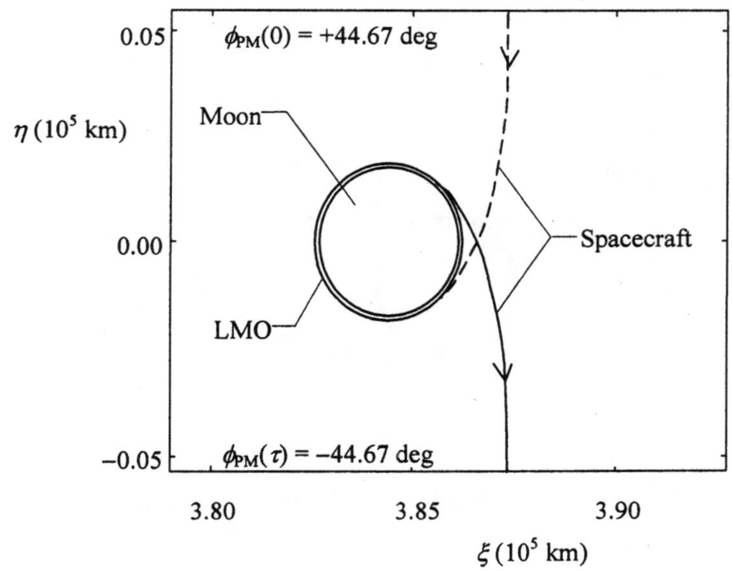

Fig. 5c. Earth-Moon-Earth trajectory, rotating coordinate system, h_{LMO} = 100 km, clockwise LMO, near-Moon space.

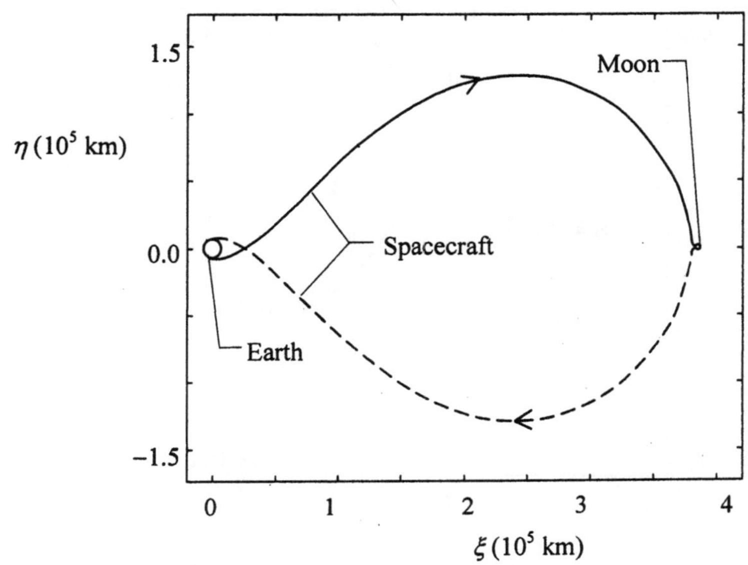

Fig. 6a. Earth-Moon-Earth trajectory, rotating coordinate system, h_{LMO} = 100 km, counterclockwise LMO, Earth-Moon space.

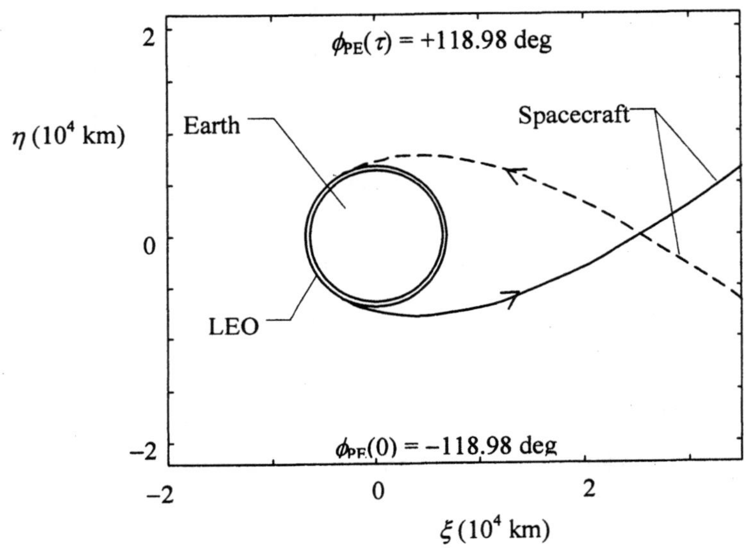

Fig. 6b. Earth-Moon-Earth trajectory, rotating coordinate system, h_{LMO} = 100 km, counterclockwise LMO, near-Earth space.

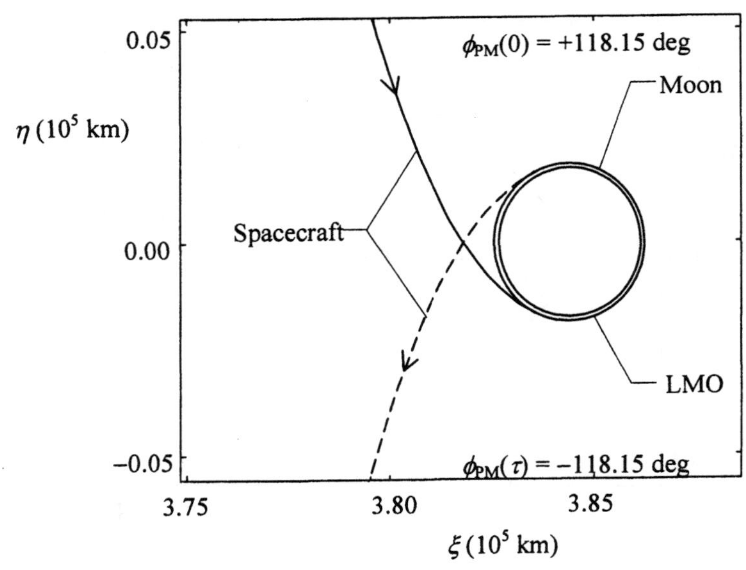

Fig. 6c. Earth-Moon-Earth trajectory, rotating coordinate system, h_{LMO} = 100 km, counterclockwise LMO, near-Moon space.

For clockwise arrival to and departure from LMO, the optimal outgoing and return trajectories are shown in Fig. 5 in Earth-Moon space, near-Earth space, and near-Moon space. Analogously, for counterclockwise arrival to and departure from LMO, the optimal outgoing and return trajectories are shown in Fig. 6 in Earth-Moon space, near-Earth space, and near-Moon space. These figures show that the optimal return trajectory is the mirror image with respect to the Earth-Moon axis of the optimal outgoing trajectory, and viceversa, once more confirming the theorem of image trajectories formulated by Miele for feasible trajectories in 1960 (Ref. 1).

6. Fixed-Time Trajectories

The results of Sections 3 and 4 show that the flight time of an optimal trajectory (4.50 days for clockwise arrival to LMO, 4.37 days for counterclockwise arrival to LMO) is considerably larger than that of the Apollo missions (2.5 to 3.2 days depending on the mission). In light of these results, the transfer problem has been solved again for a fixed flight time τ, smaller or larger than the optimal flight time.

If τ is fixed, the number of parameters to be optimized reduces to $n = 3$, namely, ΔV_{LEO}, ΔV_{LMO}, $\theta_{PE}(0)$ for an outgoing trajectory and ΔV_{LMO}, ΔV_{LEO}, $\theta_{PM}(0)$ for a return trajectory. On the other hand, the number of final conditions is still $q = 3$, namely: the radius condition, circularization condition, and tangency condition. This being the case, we are no longer in the presence of an optimization problem, but of a simple feasibility problem, which can be solved for example with the modified quasilinearization algorithm (MQA, Ref. 17). Alternatively, if SGRA is employed (Ref. 16), the restoration phase of the algorithm alone yields the solution.

6.1. Feasibility Problem. The feasibility problem is now solved for the following LEO and LMO data:

$$h_{LEO} = 463 \text{ km}, \quad r_{LEO} = 6841 \text{ km}, \quad V_{LEO} = 7.633 \text{ km/s}, \tag{23a}$$

$$h_{LMO} = 100 \text{ km}, \quad r_{LMO} = 1838 \text{ km}, \quad V_{LMO} = 1.633 \text{ km/s}, \tag{23b}$$

and these flight times:

$$\tau = 1.0, 2.0, 3.0, 4.0, 5.0, 6.0 \text{ days}. \tag{23c}$$

For LEO-to-LMO flight, the constraints are Eqs. (13b) and any of the τ values (23c). For LMO-to-LEO flight, the constraints are Eqs. (22b) and any of the τ values (23c). The unknowns include the state variables $x_P(t)$, $y_P(t)$, $u_P(t)$, $w_P(t)$ and the parameters ΔV_{LEO}, ΔV_{LMO}, $\theta_{PE}(0)$ for LEO-to-LMO flight or the parameters ΔV_{LMO}, ΔV_{LEO}, $\theta_{PM}(0)$ for LMO-to-LEO flight.

6.2. Results. The results obtained for LEO-to-LMO flight and LMO-to-LEO flight are presented in Tables 5-6. For LEO-to-LMO flight, Table 5 refers to clockwise LMO arrival; for LMO-to-LEO flight, Table 6 refers to clockwise LMO departure. Major comments are as follows:

(i) if the prescribed flight time is within one day of the optimal time, the penalty in characteristic velocity is relatively small;

(ii) if the prescribed flight time is greater than the optimal time by more than one day, the penalty in characteristic velocity becomes more severe;

(iii) if the prescribed flight time is greater than the optimal time by more than two days, no feasible trajectory exists for the given boundary conditions;

(iv) for given flight time, the outgoing and return trajectories are mirror images of one another with respect to the Earth-Moon axis, thus confirming again the theorem of image trajectories (Ref. 1).

7. Conclusions

We present a systematic study of optimal trajectories for Earth-Moon flight under the following scenario: A spacecraft initially in a counterclockwise low Earth orbit (LEO) at Space Station altitude must be transferred to either a clockwise or counterclockwise low Moon orbit (LMO) at various altitudes over the Moon surface. We study a

Table 5a. Earth-Moon flight, fixed flight time, h_{LMO} = 100 km, clockwise LMO arrival, major parameters.

τ [days]	ΔV [km/s]	ΔV_{LEO} [km/s]	ΔV_{LMO} [km/s]	$\theta_{PE}(0)$ [deg]
1.0	6.804	3.780	3.024	−127.53
2.0	4.425	3.176	1.249	−136.55
3.0	3.987	3.088	0.899	−131.53
4.0	3.892	3.070	0.822	−122.15
5.0	3.882	3.068	0.814	−111.50
6.0	3.894	3.070	0.824	−100.36

Table 5b. Earth-Moon flight, fixed flight time, h_{LMO} = 100 km, clockwise LMO arrival, phase angles at departure.

τ [days]	$\theta_{PE}(0)$ [deg]	$\theta_{M}(0)$ [deg]	$\phi_{PE}(0)$ [deg]	Note
1.0	−127.53	0.00	−127.53	See Eqs. (21)
2.0	−136.55	0.00	−136.55	
3.0	−131.53	0.00	−131.53	
4.0	−122.15	0.00	−122.15	
5.0	−111.50	0.00	−111.50	
6.0	−100.36	0.00	−100.36	

complementary problem for Moon-Earth flight with counterclockwise return to a low Earth orbit.

The assumed physical model is a simplified version of the restricted three-body problem. The optimization criterion is the total characteristic velocity and the parameters being optimized are four: initial phase angle of the spacecraft with respect to either Earth (outgoing trip) or Moon (return trip), flight time, velocity impulse at departure, velocity impulse on arrival.

Major results for both the outgoing and return trips are as follows:

Table 5c. Earth-Moon flight, fixed flight time, h_{LMO} = 100 km, clockwise LMO arrival, phase angles at arrival.

τ [days]	$\theta_{PM}(\tau)$ [deg]	$\theta_M(\tau)$ [deg]	$\phi_{PM}(\tau)$ [deg]	Note
1.0	83.79	13.11	70.68	See Eqs. (21)
2.0	60.43	26.23	34.20	
3.0	34.64	39.34	– 4.70	
4.0	18.22	52.45	–34.23	
5.0	12.90	65.57	–52.67	
6.0	14.98	78.68	–63.70	

Table 6a. Moon-Earth flight, fixed flight time, h_{LMO} = 100 km, clockwise LMO departure, major parameters.

τ [days]	ΔV [km/s]	ΔV_{LEO} [km/s]	ΔV_{LMO} [km/s]	$\theta_{PM}(0)$ [deg]
1.0	6.804	3.780	3.024	–70.68
2.0	4.425	3.176	1.249	–34.20
3.0	3.987	3.088	0.899	4.70
4.0	3.892	3.070	0.822	34.23
5.0	3.882	3.068	0.814	52.67
6.0	3.894	3.070	0.824	63.70

(i) the velocity impulse at LEO is nearly independent of the LMO altitude (see Ref. 18);
(ii) the velocity impulse at LMO decreases as the LMO altitude increases (see Ref. 18);
(iii) the flight time of an optimal trajectory is considerably larger than that of an Apollo trajectory, regardless of whether the LMO arrival/departure is clockwise or counterclockwise;
(iv) the optimal trajectories with counterclockwise LMO arrival/departure are slightly superior to the optimal trajectories with clockwise

LMO arrival/departure in terms of both characteristic velocity and flight time.

In light of (iii), a further parametric study has been performed for both the outgoing and return trips. The transfer problem has been solved again for a fixed flight time. Major results are as follows:

(v) if the prescribed flight time is within one day of the optimal flight time, the penalty in characteristic velocity is relatively small;

(vi) for larger time deviations, the penalty in characteristic velocity becomes more severe;

(vii) if the prescribed flight time is greater than the optimal time by more than two days, no feasible trajectory exists for the given boundary conditions.

While the present study has been made in inertial coordinates, conversion of the results into rotating coordinates leads to one of the most interesting findings of this paper, namely:

(viii) the optimal LEO-to-LMO trajectories and the optimal LMO-to-LEO trajectories are mirror images of one another with respect to the Earth-Moon axis;

(ix) the above result extends to optimal trajectories the theorem of image trajectory formulated by Miele for feasible trajectories in 1960 (Ref. 1).

Table 6b. Moon-Earth flight, fixed flight time, h_{LMO} = 100 km, clockwise LMO departure, phase angles at departure.

τ [days]	$\theta_{PM}(0)$ [deg]	$\theta_M(0)$ [deg]	$\phi_{PM}(0)$ [deg]	Note
1.0	–70.68	0.00	–70.68	See Eqs. (21)
2.0	–34.20	0.00	–34.20	
3.0	4.70	0.00	4.70	
4.0	34.23	0.00	34.23	
5.0	52.67	0.00	52.67	
6.0	63.70	0.00	63.70	

Table 6c. Moon-Earth flight, fixed flight time, h_{LMO} = 100 km, clockwise LMO departure, phase angles at arrival.

τ [days]	$\theta_{PE}(\tau)$ [deg]	$\theta_M(\tau)$ [deg]	$\phi_{PE}(\tau)$ [deg]	Note
1.0	140.64	13.11	127.53	See Eqs. (21)
2.0	162.77	26.22	136.55	
3.0	170.87	39.34	131.53	
4.0	174.60	52.45	122.15	
5.0	177.06	65.56	111.50	
6.0	179.04	78.68	100.36	

References

1. MIELE, A., *Theorem of Image Trajectories in the Earth-Moon Space*, Astronautica Acta, Vol. 6, No. 5, pp. 225-232, 1960.
2. MICKELWAIT, A. B., and BOOTON, R. C., *Analytical and Numerical Studies of Three-Dimensional Trajectories to the Moon*, Journal of the Aerospace Sciences, Vol. 27, No. 8, pp. 561-573, 1960.
3. CLARKE, V. C., *Design of Lunar and Interplanetary Ascent Trajectories*, AIAA Journal, Vol. 5, No. 7, pp. 1559-1567, 1963.
4. REICH, H., *General Characteristics of the Launch Window for Orbital Launch to the Moon*, Celestial Mechanics and Astrodynamics, Edited by V. G. Szebehely, Vol. 14, pp. 341-375, 1964.
5. DALLAS, C. S., *Moon-to-Earth Trajectories*, Celestial Mechanics and Astrodynamics, Edited by V. G. Szebehely, Vol. 14, pp. 391-438, 1964.
6. BAZHINOV, I. K., *Analysis of Flight Trajectories to Moon, Mars, and Venus*, Post-Apollo Space Exploration, Edited by F. Narin, Advances in the Astronautical Sciences, Vol. 20, pp. 1173-1188, 1966.
7. SHAIKH, N. A., *A New Perturbation Method for Computing Earth-Moon Trajectories*, Astronautica Acta, Vol. 12, No. 3, pp. 207-211, 1966.

8. ROSENBAUM, R., WILLWERTH, A. C., and CHUCK, W., *Powered Flight Trajectory Optimization for Lunar and Interplanetary Transfer*, Astronautica Acta, Vol. 12, No. 2, pp. 159-168, 1966.

9. MINER, W. E., and ANDRUS, J. F., *Necessary Conditions for Optimal Lunar Trajectories with Discontinuous State Variables and Intermediate Point Constraints*, AIAA Journal, Vol. 6, No. 11, pp. 2154-2159, 1968.

10. D'AMARIO, L. A., and EDELBAUM, T. N., *Minimum Impulse Three-Body Trajectories*, AIAA Journal, Vol. 12, No. 4, pp. 455-462, 1974.

11. PU, C. L., and EDELBAUM, T. N., *Four-Body Trajectory Optimization*, AIAA Journal, Vol. 13, No. 3, pp. 333-336, 1975.

12. KLUEVER, C. A., and PIERSON, B. L., *Optimal Low-Thrust Earth-Moon Transfers with a Switching Function Structure*, Journal of the Astronautical Sciences, Vol. 42, No. 3, pp. 269-283, 1994.

13. RIVAS, M. L., and PIERSON, B. L., *Dynamic Boundary Evaluation Method for Approximate Optimal Lunar Trajectories*, Journal of Guidance, Control, and Dynamics, Vol. 19, No. 4, pp. 976-978, 1996.

14. KLUEVER, C. A., and PIERSON, B. L., *Optimal Earth-Moon Trajectories Using Nuclear Electric Propulsion*, Journal of Guidance, Control, and Dynamics, Vol. 20, No. 2, pp. 239-245, 1997.

15. KLUEVER, C. A., *Optimal Earth-Moon Trajectories Using Combined Chemical-Electric Propulsion*, Journal of Guidance, Control, and Dynamics, Vol. 20, No. 2, pp. 253-258, 1997.

16. MIELE, A., HUANG, H. Y., and HEIDEMAN, J. C., *Sequential Gradient-Restoration Algorithm for the Minimization of Constrained Functions: Ordinary and Conjugate Gradient Versions*, Journal of Optimization Theory and Applications, Vol. 4, No. 4, pp. 213-243, 1969.

17. MIELE, A., NAQVI, S., LEVY, A. V., and IYER, R. R., *Numerical Solutions of Nonlinear Equations and Nonlinear Two-*

Point Boundary-Value Problems, Advances in Control Systems, Edited by C. T. Leondes, Academic Press, New York, New York, Vol. 8, pp. 189-215, 1971.

18. MIELE, A. and MANCUSO, S., *Optimal Trajectories for Earth-Moon-Earth Flight*, Aero-Astronautics Report 295, Rice University, 1998.

3

Design of Mars Missions

A. MIELE[1] AND T. WANG[2]

Abstract. This paper deals with the optimal design of round-trip Mars missions, starting from LEO (low Earth orbit), arriving to LMO (low Mars orbit), and then returning to LEO after a waiting time in LMO.

The assumed physical model is the restricted four-body model, including Sun, Earth, Mars, and spacecraft. The optimization problem is formulated as a mathematical programming problem: the total characteristic velocity (the sum of the velocity impulses at LEO and LMO) is minimized, subject to the system equations and boundary conditions of the restricted four-body model. The mathematical programming problem is solved via the sequential gradient-restoration algorithm employed in conjunction with a variable-stepsize integration technique to overcome the numerical difficulties due to large changes in the gravity field near Earth and near Mars.

The results lead to a baseline optimal trajectory computed under the assumption that the Earth and Mars orbits around Sun are circular and coplanar. The baseline optimal trajectory resembles a Hohmann transfer trajectory, but is not a Hohmann transfer trajectory, owing to the disturbing influence exerted by Earth/Mars on the terminal branches of the trajectory. For the baseline optimal trajectory, the total characteristic velocity of a round-trip Mars

[1] Research Professor and Foyt Professor Emeritus of Engineering, Aerospace Sciences, and Mathematical Sciences, Aero-Astronautics Group, Rice University, Houston, Texas 77005-1892, USA.

[2] Senior Research Scientist, Aero-Astronautics Group, Rice University, Houston, Texas 77005-1892, USA.

mission is 11.30 km/s (5.65 km/s each way) and the total mission time is 970 days (258 days each way plus 454 days waiting in LMO).

An important property of the baseline optimal trajectory is the asymptotic parallelism property: For optimal transfer, the spacecraft inertial velocity must be parallel to the inertial velocity of the closest planet (Earth or Mars) at the entrance to and exit from deep interplanetary space. For both the outgoing and return trips, asymptotic parallelism occurs at the end of the first day and at the beginning of the last day. Another property of the baseline optimal trajectory is the near-mirror property. The return trajectory can be obtained from the outgoing trajectory via a sequential procedure of rotation, reflection, and inversion.

Departure window trajectories are next-to-best trajectories. They are suboptimal trajectories obtained by changing the departure date, hence changing the Mars/Earth inertial phase angle difference at departure. For the departure window trajectories, the asymptotic parallelism property no longer holds in the departure branch, but still holds in the arrival branch. On the other hand, the near-mirror property no longer holds.

Key Words. Flight mechanics, astrodynamics, celestial mechanics, Earth-to-Mars missions, round-trip Mars missions, mirror property, asymptotic parallelism property, optimization, sequential gradient restoration algorithm.

1. Introduction

Several years ago, a research program dealing with the optimization and guidance of flight trajectories from Earth to Mars and back was initiated at Rice University. The decision was based on the recognition that the involvement of the USA with the Mars problem had been growing in recent years and it can be expected to grow in the foreseeable future (Refs. 1-15). Our feeling was that, in attacking the Mars problem, we should start with simple models and then go to models of increasing complexity. Accordingly, this paper deals with the preliminary results obtained with a relatively simple model, yet sufficiently realistic to capture some of the essential elements of the flight from Earth to Mars and back (Refs. 16-19).

1.1. Mission Alternatives, Types, Objectives. There are two basic alternatives for Mars missions: robotic missions and manned missions, the latter being considerably more complex than the former. Within each alternative, we can distinguish two types of missions: exploratory (survey) missions and sample taking (sample return) missions.

Regardless of alternative and type, there is a basic maneuver which is common to every Mars mission, namely, the transfer of a spacecraft from a low Earth orbit (LEO) to a low Mars orbit (LMO) and back. For both LEO-to-LMO transfer and LMO-to-LEO transfer, the first objective is to contain the propellant assumption; the second objective is to contain the flight time, if at all possible.

1.2. Characteristic Velocity. Under certain conditions, the propellant consumption is monotonically related to the so-called characteristic velocity, the sum of the velocity impulses applied to the spacecraft via rocket engines. In turn, by definition, each velocity impulse is a positive quantity, regardless of whether its action is accelerating or decelerating, in-plane or out-of-plane.

In astrodynamics, it is customary to replace the consideration of propellant consumption with the consideration of characteristic velocity, with the following advantage: the characteristic velocity is independent of the spacecraft structural factor and engine specific impulse, while this is not the case with the propellant consumption. Indeed, the characteristic velocity truly "characterizes" the mission itself.

1.3. Optimal Trajectories. This presentation is centered on the study of the optimal trajectories, namely, trajectories minimizing the characteristic velocity. This study is important in that it provides the basis for the development of guidance schemes approximating the optimal trajectories in real time. In turn, this requires the knowledge of some fundamental, albeit easily implementable property of the optimal trajectories. This is precisely the case with the asymptotic parallelism condition at the entrance to and exit from deep interplanetary space: For both the outgoing and return trips, minimization of the characteristic velocity is achieved if the spacecraft inertial velocity is parallel to the inertial velocity of the closest planet (Earth or Mars) at the entrance to and exit from deep interplanetary space.

2. Four-Body Model

At every point of the trajectory, the spacecraft is subject to the gravitational attractions of Earth, Mars, and Sun. Therefore, we are in the presence of a four-body problem, the four bodies being the spacecraft, Earth, Mars, and Sun (Fig. 1a). Assuming that the Sun is fixed in space, the complete four-body model is described by 18 nonlinear ordinary differential equations (ODEs) in the three-dimensional case and by 12 nonlinear ODEs in the two-dimensional case (planar case). Two possible simplifications are described below.

2.1. Patched Conics Model. This model consists in subdividing an Earth-to-Mars trajectory into three segments: a near-Earth segment in which Earth gravity is dominant; a deep interplanetary space segment in which Sun gravity is dominant; a near-Mars segment in which Mars gravity is dominant. Under this scenario, the four-body problem is replaced by a succession of two-body problems, each described in the planar case by four ODEs, for which analytical solutions are available.

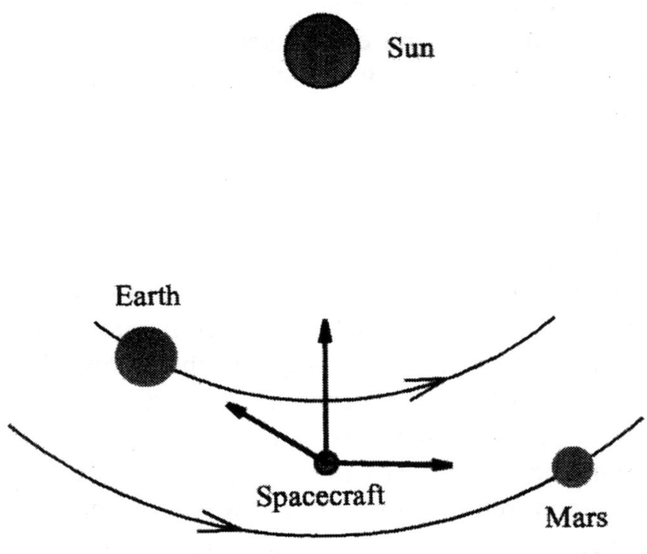

Fig. 1a. Four body-problem: gravitational forces.

Then, the segmented solutions must be patched together in such a way that some continuity conditions are satisfied at the interface between contiguous segments.

Even though the method of patched conics has been widely used in the literature, our experience with it has been rather disappointing for the reason indicated below. Near the interface between contiguous segments, there is a small region in which two of the three gravitational attractions are of the same order. Neglecting one of them on each side of the interface induces small local errors in the spacecraft acceleration, which in turn induce large errors in velocity and position owing to long integration times. In light of this statement, we discarded the patched conics model, replacing it with the restricted four-body model.

2.2. Restricted Four-Body Model. This model consists in assuming that the inertial motions of Earth and Mars are determined only by Sun, while the inertial motion of the spacecraft is determined by Earth, Mars, and Sun. In the planar case, this is equivalent to splitting the complete system of order 12 into three subsystems, each of order four: the Earth, Mars, and spacecraft subsystems. The first two subsystems can be integrated independently of the third; the third subsystem can be integrated once the solutions of the first two are known. This is the essential simplification provided by the restricted four-body model, while avoiding the pitfalls of the patched conics model.

3. System Description

Let LEO denote a low Earth orbit, and let LMO denote a low Mars orbit. We study the LEO-to-LMO transfer [LMO-to-LEO transfer] of a spacecraft under the following scenario (Fig. 1b). Initially, the spacecraft moves in a circular orbit around Earth [Mars]; an accelerating velocity impulse is applied tangentially to LEO [LMO], and its magnitude is such that the spacecraft escapes from near-Earth [near-Mars] space into deep interplanetary space. Then, the spacecraft takes a long journey along an interplanetary orbit around the Sun, enters near-Mars [near-Earth] space, and reaches tangentially the low Mars orbit [low Earth orbit]. Here, a decelerating velocity impulse is applied tangentially to LMO [LEO] so as to achieve circularization of the motion around Mars [Earth].

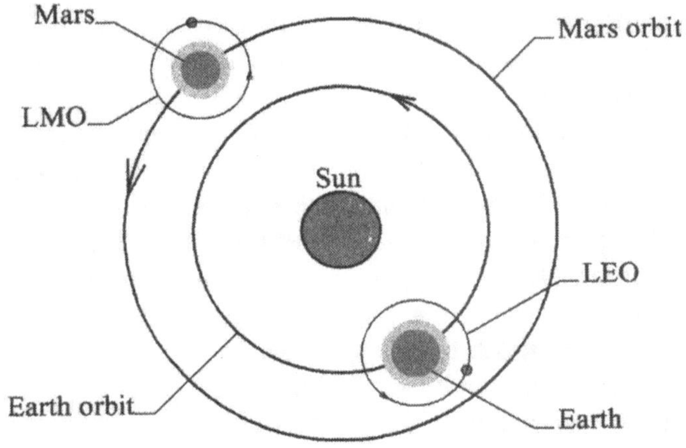

Fig. 1b. Earth and Mars orbits around Sun and spacecraft terminal orbits.

The following hypotheses are employed: (A1) the Sun is fixed in space; (A2) Earth and Mars are subject only to the Sun gravity; (A3) the eccentricity of the Earth and Mars orbits around the Sun is neglected, implying circular planetary motions; (A4) the inclination of the Mars orbital plane vis-à-vis the Earth orbital plane is neglected, implying planar spacecraft motion; (A5) the spacecraft is subject to the gravitational attractions of Earth, Mars, and Sun along the entire trajectory; (A6) for the outgoing and return trips, the class of two-impulse trajectories is considered, with the impulses being applied at the terminal points of the trajectories; (A7) for the outgoing and return trips, circularization of motion around the relevant planet is assumed both before departure and after arrival.

Having adopted the restricted four-body model to achieve increased precision with respect to the patched conics model, we are simultaneously interested in five motions: the inertial motions of Earth, Mars, and spacecraft with respect to the Sun; the relative motions of the spacecraft with respect to Earth and Mars. To study these motions, we employ three coordinate systems: Sun coordinate system (SCS), Earth coordinate system (ECS), and Mars coordinate system (MCS).

SCS is an inertial coordinate system; its origin is the Sun center and its axes x, y are fixed in space; in particular, the x-axis points to the initial position of the Earth center and the y-axis is orthogonal to the x-axis. ECS is a relative-to-Earth coordinate system; its origin is the Earth center and its axes x', y' are parallel to the axes x, y of the Sun coordinate system. MCS is a relative-to-Mars coordinate system; its origin is the Mars center and its axes x'', y'' are parallel to the axes x, y of the Sun coordinate system.

Clearly, ECS and MCS translate without rotation w.r.t. SCS. Their origins E and M move around the Sun with constant angular velocities ω_E and ω_M. The angular velocity difference $\omega_E - \omega_M$ is also constant.

In this paper, the inertial motions of the spacecraft, Earth, and Mars are described in Sun coordinates, while the spacecraft boundary conditions are described in relative-to-planet coordinates. If polar coordinates are used, a position vector is defined via the radial distance r and phase angle θ, while a velocity vector is defined via the velocity modulus V and local path inclination γ. If Cartesian coordinates are used, a position vector is defined its via components x, y and a velocity vector via its components u, w.

Let E, M, S denote the centers of Earth, Mars, and Sun; let μ_E, μ_M, μ_S denote the gravitational constants of Earth, Mars, and Sun; let P denote the spacecraft; let t denote the time, $0 \leq t \leq \tau$, with 0 the initial time and τ the final time. Below, we give the system equations for Earth, Mars, and spacecraft in Sun coordinates; for details, see Refs. 16-19.

3.1. Earth. Subject to the Sun gravitational attraction and neglecting the orbital eccentricity, we approximate the Earth (subscript E) trajectory around the Sun with a circle. Hence, in polar coordinates, the position and velocity of Earth are given by

$$\text{(SCS)} \qquad r_E = r_E(0) , \tag{1a}$$

$$\theta_E = t\sqrt{(\mu_S / r_E^3)} + \theta_E(0) , \tag{1b}$$

$$V_E = \sqrt{(\mu_S / r_E)} , \tag{1c}$$

$$\gamma_E = 0 . \tag{1d}$$

In Cartesian coordinates, the position and velocity of Earth are described by

$$\text{(SCS)} \qquad x_E = r_E \cos\theta_E , \tag{2a}$$

$$y_E = r_E \sin\theta_E , \tag{2b}$$

$$u_E = -V_E \sin\theta_E , \tag{2c}$$

$$w_E = +V_E \cos\theta_E , \tag{2d}$$

with

$$\text{(SCS)} \qquad r_E = \sqrt{(x_E^2 + y_E^2)}, \tag{3a}$$

$$V_E = \sqrt{(u_E^2 + w_E^2)}, \tag{3b}$$

$$x_E u_E + y_E w_E = 0. \tag{3c}$$

Equation (3c) is an orthogonality condition between vec(SE) and vec(V_E), where vec stands for vector.

3.2. Mars. Subject to the Sun gravitational attraction and neglecting the orbital eccentricity, we approximate the Mars (subscript M) trajectory around the Sun with a circle. Hence, in polar coordinates, the position and velocity of Mars are given by

$$\text{(SCS)} \qquad r_M = r_M(0) , \tag{4a}$$

$$\theta_M = t\sqrt{(\mu_S / r_M^3)} + \theta_M(0) , \tag{4b}$$

$$V_M = \sqrt{(\mu_S / r_M)} , \tag{4c}$$

$$\gamma_M = 0 . \tag{4d}$$

In Cartesian coordinates, the position and velocity of Mars are described by

$$\text{(SCS)} \qquad x_M = r_M \cos \theta_M , \tag{5a}$$

$$y_M = r_M \sin \theta_M , \tag{5b}$$

$$u_M = - V_M \sin \theta_M , \tag{5c}$$

$$w_M = +V_M \cos \theta_M , \tag{5d}$$

with

$$\text{(SCS)} \qquad r_M = \sqrt{(x_M^2 + y_M^2)}, \tag{6a}$$

$$V_M = \sqrt{(u_M^2 + w_M^2)}, \tag{6b}$$

$$x_M u_M + y_M w_M = 0. \tag{6c}$$

Equation (6c) is an orthogonality condition between vec(SM) and vec(V_M), where vec stands for vector.

3.3. Spacecraft. Subject to the gravitational attractions of Sun, Earth, and Mars along the entire trajectory, the motion of the spacecraft (subscript P) around the Sun is described by the following differential equations in the coordinates x_P, y_P of the position vector and the components u_P, w_P of the velocity vector:

$$\text{(SCS)} \qquad dx_P/dt = u_P , \tag{7a}$$

$$dy_P/dt = w_P , \tag{7b}$$

$$du_P/dt = - \mu_S x_P / r_P^3 - \mu_E (x_P - x_E)/ r_{PE}^3$$

$$- \mu_M (x_P - x_M)/ r_{PM}^3 , \tag{7c}$$

$$dw_P/dt = -\mu_S y_P/r_P^3 - \mu_E(y_P - y_E)/r_{PE}^3 - \mu_M(y_P - y_M)/r_{PM}^3. \quad (7d)$$

Here r_P, r_{PE}, r_{PM} are the radial distances of the spacecraft from the Sun, Earth, and Mars; these quantities can be computed via the relations

$$\text{(SCS)} \quad r_P = \sqrt{(x_P^2 + y_P^2)}, \quad (8a)$$

$$r_{PE} = \sqrt{[(x_P - x_E)^2 + (y_P - y_E)^2]}, \quad (8b)$$

$$r_{PM} = \sqrt{[(x_P - x_M)^2 + (y_P - y_M)^2]}. \quad (8c)$$

4. Boundary Conditions

4.1. Outgoing Trip, Departure. In polar coordinates, the spacecraft conditions at the departure from LEO (time $t = 0$) are given by

$$\text{(ECS)} \quad r_{PE}(0) = r_{LEO}, \quad (9a)$$

$$\theta_{PE}(0) = \text{free}, \quad (9b)$$

$$V_{PE}(0) = V_{LEO} + \Delta V_{LEO}$$

$$= \sqrt{(\mu_E / r_{LEO})} + \Delta V_{LEO}, \quad (9c)$$

$$\gamma_{PE}(0) = 0. \quad (9d)$$

Relative to Earth r_{PE}, θ_{PE}, V_{PE}, γ_{PE} are the radial distance, phase angle, velocity, and path inclination of the spacecraft; V_{LEO} is the spacecraft velocity in the low Earth orbit prior to application of the tangential, accelerating velocity impulse; ΔV_{LEO} is the accelerating velocity impulse at LEO; $V_{PE}(0)$ is the spacecraft velocity after application of the accelerating velocity impulse.

The corresponding equations in Cartesian coordinates are

$$\text{(ECS)} \quad x_{PE}(0) = r_{PE}(0) \cos\theta_{PE}(0), \tag{10a}$$

$$y_{PE}(0) = r_{PE}(0) \sin\theta_{PE}(0), \tag{10b}$$

$$u_{PE}(0) = - V_{PE}(0) \sin\theta_{PE}(0), \tag{10c}$$

$$w_{PE}(0) = +V_{PE}(0) \cos\theta_{PE}(0), \tag{10d}$$

with

$$\text{(ECS)} \quad r_{PE}(0) = \sqrt{[x_{PE}^2(0) + y_{PE}^2(0)]}, \tag{11a}$$

$$V_{PE}(0) = \sqrt{[u_{PE}^2(0) + w_{PE}^2(0)]}, \tag{11b}$$

$$x_{PE}(0)\, u_{PE}(0) + y_{PE}(0)\, w_{PE}(0) = 0. \tag{11c}$$

Equation (11c) is an orthogonality condition between vec(EP(0)) and vec($V_{PE}(0)$), meaning that the accelerating velocity impulse ΔV_{LEO} is tangential to LEO.

4.2. Outgoing Trip, Arrival. In polar coordinates, the spacecraft conditions at the arrival to LMO (time t = τ) are given by

$$\text{(MCS)} \quad r_{PM}(\tau) = r_{LMO}, \tag{12a}$$

$$\theta_{PM}(\tau) = \text{free}, \tag{12b}$$

$$\begin{aligned} V_{PM}(\tau) &= V_{LMO} + \Delta V_{LMO} \\ &= \sqrt{(\mu_M / r_{LMO})} + \Delta V_{LMO}, \end{aligned} \tag{12c}$$

$$\gamma_{PM}(\tau) = 0. \tag{12d}$$

Relative to Mars r_{PM}, θ_{PM}, V_{PM}, γ_{PM} are the radial distance, phase angle, velocity, and path inclination of the spacecraft; V_{LMO} is the spacecraft velocity in the low Mars orbit after application of the tangential, decelerating velocity impulse; ΔV_{LMO} is the decelerating velocity impulse at LMO; $V_{PM}(\tau)$ is the spacecraft velocity before application of the decelerating velocity impulse.

The corresponding equations in Cartesian coordinates are

$$\text{(MCS)} \quad x_{PM}(\tau) = r_{PM}(\tau)\cos\theta_{PM}(\tau), \tag{13a}$$

$$y_{PM}(\tau) = r_{PM}(\tau)\sin\theta_{PM}(\tau), \tag{13b}$$

$$u_{PM}(\tau) = -V_{PM}(\tau)\sin\theta_{PM}(\tau), \tag{13c}$$

$$w_{PM}(\tau) = +V_{PM}(\tau)\cos\theta_{PM}(\tau), \tag{13d}$$

with

$$\text{(MCS)} \quad r_{PM}(\tau) = \surd\,[x_{PM}^2(\tau) + y_{PM}^2(\tau)], \tag{14a}$$

$$V_{PM}(\tau) = \surd\,[u_{PM}^2(\tau) + w_{PM}^2(\tau)], \tag{14b}$$

$$x_{PM}(\tau)\,u_{PM}(\tau) + y_{PM}(\tau)\,w_{PM}(\tau) = 0. \tag{14c}$$

Equation (14c) is an orthogonality condition between vec(MP(τ)) and vec($V_{PM}(\tau)$), meaning that the decelerating velocity impulse ΔV_{LMO} is tangential to LMO.

4.3. Return Trip, Departure. In polar coordinates, the spacecraft conditions at the departure from LMO (time $t = 0$) are given by

$$\text{(MCS)} \quad r_{PM}(0) = r_{LMO}, \tag{15a}$$

$$\theta_{PM}(0) = \text{free}, \tag{15b}$$

$$V_{PM}(0) = V_{LMO} + \Delta V_{LMO} = \surd(\mu_M / r_{LMO}) + \Delta V_{LMO}, \tag{15c}$$

$$\gamma_{PM}(0) = 0 \,. \tag{15d}$$

Formally, Eqs. (15) can be obtained from Eqs. (12) by simply replacing the time $t = \tau$ with the time $t = 0$. However, there is a difference of interpretation: V_{LMO} is now the spacecraft velocity in the low Mars orbit before application of the tangential, accelerating velocity impulse; ΔV_{LMO} is the accelerating velocity impulse at LMO; $V_{PM}(0)$ is the spacecraft velocity after application of the accelerating velocity impulse.

The corresponding equations in Cartesian coordinates are

$$\text{(MCS)} \quad x_{PM}(0) = r_{PM}(0) \cos\theta_{PM}(0)\,, \tag{16a}$$

$$y_{PM}(0) = r_{PM}(0) \sin\theta_{PM}(0)\,, \tag{16b}$$

$$u_{PM}(0) = -\, V_{PM}(0) \sin\theta_{PM}(0)\,, \tag{16c}$$

$$w_{PM}(0) = +V_{PM}(0) \cos\theta_{PM}(0)\,, \tag{16d}$$

with

$$\text{(MCS)} \quad r_{PM}(0) = \sqrt{\,[x^2_{PM}(0) + y^2_{PM}(0)]}, \tag{17a}$$

$$V_{PM}(0) = \sqrt{\,[u^2_{PM}(0) + w^2_{PM}(0)]}, \tag{17b}$$

$$x_{PM}(0)\, u_{PM}(0) + y_{PM}(0)\, w_{PM}(0) = 0. \tag{17c}$$

Equation (17c) is an orthogonality condition between vec(MP(0)) and vec($V_{PM}(0)$), meaning that the accelerating velocity impulse ΔV_{LMO} is tangential to LMO.

4.4. Return Trip, Arrival. In polar coordinates, the spacecraft conditions at the arrival to LEO (time $t = \tau$) are given by

$$\text{(ECS)} \quad r_{PE}(\tau) = r_{LEO}\,, \tag{18a}$$

$$\theta_{PE}(\tau) = \text{free}\,, \tag{18b}$$

$$\begin{aligned} V_{PE}(\tau) &= V_{LEO} + \Delta V_{LEO} \\ &= \sqrt{(\mu_E / r_{LEO})} + \Delta V_{LEO}\,, \end{aligned} \tag{18c}$$

$$\gamma_{PE}(\tau) = 0\,. \tag{18d}$$

Formally, Eqs. (18) can be obtained from Eqs. (9) by simply replacing the time $t = 0$ with the time $t = \tau$. However, there is a difference of interpretation: V_{LEO} is now the spacecraft velocity in the low Earth orbit after application of the tangential, decelerating velocity impulse; ΔV_{LEO} is the decelerating velocity impulse at LEO; $V_{PE}(\tau)$ is the spacecraft velocity before application of the decelerating velocity impulse.

The corresponding equations in Cartesian coordinates are

$$\text{(ECS)} \qquad x_{PE}(\tau) = r_{PE}(\tau)\cos\theta_{PE}(\tau)\,, \tag{19a}$$

$$y_{PE}(\tau) = r_{PE}(\tau)\sin\theta_{PE}(\tau)\,, \tag{19b}$$

$$u_{PE}(\tau) = -\,V_{PE}(\tau)\sin\theta_{PE}(\tau)\,, \tag{19c}$$

$$w_{PE}(\tau) = +V_{PE}(\tau)\cos\theta_{PE}(\tau)\,, \tag{19d}$$

with

$$\text{(ECS)} \qquad r_{PE}(\tau) = \sqrt{[x_{PE}^2(\tau) + y_{PE}^2(\tau)]}, \tag{20a}$$

$$V_{PE}(\tau) = \sqrt{[u_{PE}^2(\tau) + w_{PE}^2(\tau)]}, \tag{20b}$$

$$x_{PE}(\tau)\,u_{PE}(\tau) + y_{PE}(\tau)\,w_{PE}(\tau) = 0. \tag{20c}$$

Equation (20c) is an orthogonality condition between vec(EP(τ)) and vec($V_{PE}(\tau)$), meaning that the decelerating velocity impulse ΔV_{LEO} is tangential to LEO.

5. Coordinate Transformations

Due to the fact that the spacecraft equations of motion are given in inertial coordinates (SCS), while the spacecraft boundary conditions are given in relative-to-planet coordinates (ECS) or (MCS), coordinate transformations are needed to pass from one system to another at the terminal points of the outgoing and return trips. The transformations are given below.

(i) ECS-to-SCS Transformation. For the outgoing trip, this transformation is to be employed to convert spacecraft conditions at the departure from LEO (time t = 0) from relative-to-Earth coordinates to inertial coordinates. In Cartesian coordinates,

$$x_P(0) = x_E(0) + x_{PE}(0), \tag{21a}$$

$$y_P(0) = y_E(0) + y_{PE}(0), \tag{21b}$$

$$u_P(0) = u_E(0) + u_{PE}(0), \tag{21c}$$

$$w_P(0) = w_E(0) + w_{PE}(0). \tag{21d}$$

(ii) SCS-to-MCS Transformation. For the outgoing trip, this transformation is to be employed to convert spacecraft conditions at the arrival to LMO (time $t = \tau$) from inertial coordinates to relative-to-Mars coordinates. In Cartesian coordinates,

$$x_{PM}(\tau) = x_P(\tau) - x_M(\tau), \tag{22a}$$

$$y_{PM}(\tau) = y_P(\tau) - y_M(\tau), \tag{22b}$$

$$u_{PM}(\tau) = u_P(\tau) - u_M(\tau), \tag{22c}$$

$$w_{PM}(\tau) = w_P(\tau) - w_M(\tau). \tag{22d}$$

(iii) MCS-to-SCS Transformation. For the return trip, this transformation is to be employed to convert spacecraft conditions at the departure from LMO (time t = 0) from relative-to-Mars coordinates to inertial coordinates. In Cartesian coordinates,

$$x_P(0) = x_M(0) + x_{PM}(0), \tag{23a}$$

$$y_P(0) = y_M(0) + y_{PM}(0), \tag{23b}$$

$$u_P(0) = u_M(0) + u_{PM}(0), \tag{23c}$$

$$w_P(0) = w_M(0) + w_{PM}(0). \tag{23d}$$

(iv) SCS-to-ECS Transformation. For the return trip, this transformation is to be employed to convert spacecraft conditions at the arrival to LEO (time $t = \tau$) from inertial coordinates to relative-to-Earth coordinates. In Cartesian coordinates,

$$x_{PE}(\tau) = x_P(\tau) - x_E(\tau), \tag{24a}$$

$$y_{PE}(\tau) = y_P(\tau) - y_E(\tau), \tag{24b}$$

$$u_{PE}(\tau) = u_P(\tau) - u_E(\tau), \tag{24c}$$

$$w_{PE}(\tau) = w_P(\tau) - w_E(\tau). \tag{24d}$$

6. Mathematical Programming Problems

In this section, we formulate the problem of the optimal round-trip trajectory as a mathematical programming problem. The complete problem can be decomposed into three separate problems to be solved in sequence: (i) determination of the optimal trajectory for the outgoing trip; (ii) determination of the optimal trajectory for the return trip; (iii) determination of the waiting time in the low Mars orbit.

6.1. Outgoing Trip. The optimization of a LEO-to-LMO transfer can be reduced to a mathematical programming problem involving the following performance index, constraints, and parameters.

Performance Index. The most obvious performance is the characteristic velocity,

$$\Delta V = \Delta V_{LEO} + \Delta V_{LMO}\,, \tag{25}$$

which is the sum of the terminal velocity impulses: ΔV_{LEO} is the accelerating velocity impulse at LEO (Earth coordinates) and ΔV_{LMO} is the decelerating velocity impulse at LMO (Mars coordinates).

Constraints. The departure conditions include the radius condition (11a), (9a), decircularization condition (11b), (9c), and tangency condition (11c) [for brevity, constraints (11)]. Satisfaction of the departure conditions is trivial for any choice of the parameters ΔV_{LEO} and $\theta_{PE}(0)$. By the same token, the differential system (7) is never violated if a forward integration is performed with SCS initial conditions consistent with (11) and (21). The only constraints to be enforced are the final conditions, which include the radius condition (14a), (12a), circularization condition (14b), (12c), and tangency condition (14c) [for brevity, constraints (14)].

Parameters. Let a, b, c denote the following vector parameters:

$$a = [r_{LEO},\, r_{LMO},\, \Delta V_{LEO},\, \Delta V_{LMO},\, \tau,\, \theta_{PE}(0),\, \Delta\theta(0)]^T, \tag{26a}$$

$$b = [r_{LEO},\, r_{LMO}]^T, \tag{26b}$$

$$c = [\Delta V_{LEO},\, \Delta V_{LMO},\, \tau,\, \theta_{PE}(0),\, \Delta\theta(0)]^T. \tag{26c}$$

The 7×1 vector a includes the major parameters governing a LEO-to-LMO trajectory; the 2×1 vector b includes the components of x that are fixed, namely, the radii of the terminal orbits r_{LEO} and r_{LMO}; the 5×1 vector c includes the components of a that must be optimized, namely, the terminal velocity impulses ΔV_{LEO} and ΔV_{LMO}, transfer time τ, spacecraft/Earth relative phase angle at departure $\theta_{PE}(0)$, and Mars/Earth inertial phase angle difference at departure $\Delta\theta(0) = \theta_M(0) - \theta_E(0)$. Note that $\Delta\theta(0) = \theta_M(0)$ if one sets $\theta_E(0) = 0$ by definition.

Problem P1. For the outgoing trip, given the vector parameter b [see (26b)], minimize the performance index (25) w.r.t. the vector parameter c [see (26c)], subject to the constraints (14).

Problem P1 is characterized by $n = 5$ variables and $q = 3$ constraints; hence, the number of degrees of freedom is $n - q = 2$, implying that there are only two independent parameters, for instance, $\theta_{PE}(0)$ and $\Delta\theta(0)$. The solution of Problem P1 is called the baseline optimal trajectory and yields the smallest value of the characteristic velocity (25) compatible with a given pair (r_{LEO}, r_{LMO}).

6.2. Return Trip. The optimization of a LMO-to-LEO transfer can be reduced to a mathematical programming problem involving the following performance index, constraints, and parameters.

Performance Index. The most obvious performance is the characteristic velocity,

$$\Delta V = \Delta V_{LMO} + \Delta V_{LEO}\,, \tag{27}$$

which is the sum of the terminal velocity impulses: ΔV_{LMO} is the accelerating velocity impulse at LMO (Mars coordinates) and ΔV_{LEO} is the decelerating velocity impulse at LEO (Earth coordinates).

Constraints. The departure conditions include the radius condition (17a), (15a), decircularization condition (17b), (15c), and tangency condition (17c) [for brevity, constraints (17)]. Satisfaction of the departure conditions is trivial for any choice of the parameters ΔV_{LMO} and $\theta_{PM}(0)$. By the same token, the differential system (7) is never violated if a forward integration is performed with SCS initial conditions consistent with (17) and (23). The only constraints to be enforced are the final conditions, which include the radius condition (20a), (18a), circularization condition (20b), (18c), and tangency condition (20c) [for brevity, constraints (20)].

Parameters. Let *a, b, c* denote the following vector parameters:

$$a = [r_{LMO}, r_{LEO}, \Delta V_{LMO}, \Delta V_{LEO}, \tau, \theta_{PM}(0), \Delta\theta(0)]^T, \tag{28a}$$

$$b = [r_{LMO}, r_{LEO}]^T, \tag{28b}$$

$$c = [\Delta V_{LMO}, \Delta V_{LEO}, \tau, \theta_{PM}(0), \Delta\theta(0)]^T. \tag{28c}$$

The 7×1 vector a includes the major parameters governing a LMO-to-LEO trajectory; the 2×1 vector b includes the components of a that are fixed, namely, the radii of the terminal orbits r_{LMO} and r_{LEO}; the 5×1 vector c includes the components of a that must be optimized, namely, the terminal velocity impulses ΔV_{LMO} and ΔV_{LEO}, transfer time τ, spacecraft/Mars relative phase angle at departure $\theta_{PM}(0)$, and Mars/Earth inertial phase angle difference at departure $\Delta\theta(0) = \theta_M(0) - \theta_E(0)$. Note that $\Delta\theta(0) = \theta_M(0)$ if one sets $\theta_E(0) = 0$ by definition.

Problem P2. For the return trip, given the vector parameter b [see (28b)], minimize the performance index (27) w.r.t. the vector parameter c [see (28c)], subject to the constraints (20).

Problem P2 is characterized by $n = 5$ variables and $q = 3$ constraints; hence, the number of degrees of freedom is $n - q = 2$, implying that there are only two independent parameters, for instance, $\theta_{PM}(0)$ and $\Delta\theta(0)$. The solution of Problem P2 is called the baseline optimal trajectory and yields the smallest value of the characteristic velocity (27) compatible with a given pair (r_{LMO}, r_{LEO}).

6.3. Waiting Time. For the outgoing trip, celestial mechanics requires that Mars be ahead of Earth at departure from LEO, but behind Earth at arrival to LMO; hence, for Problem P1, $\Delta\theta(0) > 0$ and $\Delta\theta(\tau) < 0$. For the return trip, celestial mechanics requires also that Mars be ahead of Earth at departure from LMO, but behind Earth at arrival to LEO; hence, for Problem P2, $\Delta\theta(0) > 0$ and $\Delta\theta(\tau) < 0$. This implies that the spacecraft cannot return immediately to Earth and is forced to wait a relatively long time in LMO to allow the Mars/Earth inertial phase angle difference to transition from the optimal arrival value of the outgoing trip to the optimal departure value of the return trip.

For the optimal trajectory, the waiting time on LMO can be computed with the relation

$$(\omega_E - \omega_M)\, \tau_{waiting} = 360 + [\Delta\theta(\tau)]_{P1} - [\Delta\theta(0)]_{P2}\,, \tag{29}$$

with angles measured in degrees and time in days.

6.4. Delay Time. Assume now that, due to technical difficulties, it is not possible to fire the rocket engines at the appropriate departure day for the return trip nor within the tolerance supplied by the departure window (see Section 10). This implies that there is a further delay time in LMO, which can be computed with the relation

$$(\omega_E - \omega_M)\, \tau_{delay} = 360\,, \tag{30}$$

with angles measured in degrees and time in days.

7. Computational Information

7.1. Algorithm. The sequential gradient-restoration algorithm (SGRA) in mathematical programming format is used to solve the mathematical programming problems of Section 6. SGRA is an iterative technique which involves a sequence of two-phase cycles, each cycle including a gradient phase and a restoration phase. In the gradient phase, the augmented performance index (performance index augmented by the constraints weighted via appropriate Lagrange multipliers) is decreased, while avoiding excessive constraint violation. In the restoration phase, the constraint error is decreased, while avoiding excessive change in the problem variables. In a complete gradient-restoration cycle, the performance index is decreased, while the constraints are satisfied to a preselected accuracy. Thus, a succession of feasible suboptimal solutions is generated, each new solution being an improvement over the previous one from the point of view of the performance index (25) or (27).

Note that SGRA is available in both mathematical programming format and optimal control format. For mathematical programming problems, SGRA was developed by Miele at al in both ordinary-gradient version and conjugate-gradient version (Ref. 20). Several variations of SGRA were also developed, but the basic form proved to be the more reliable, because of its robustness and stability properties (Ref. 21). For optimal control problems, the development of SGRA by Miele at al has been parallel to that for mathematical programming problems; see Refs. 22-24 for early versions and Refs. 25-27 for recent versions. Also for optimal control problems, an industrial version of SGRA has been developed by McDonnell-Douglas Technical Service Company under the code name SEGRAM (Ref. 28) and is being used at NASA-Johnson Space Center.

7.2. Integration Scheme. The achievement of constraint satisfaction and optimality condition satisfaction requires multiple integrations of the system equations of the restricted four-body model. The integration process is computationally expensive and it is difficult to achieve the desired accuracy, owing to the fact that the total gravitational acceleration changes rapidly in near-Earth space and near-Mars space, but slowly in deep interplanetary space. Indeed, orbital periods are of order one hour if the Earth gravity or Mars gravity is dominant, but of order one year if the Sun gravity is dominant. The above difficulties can be overcome by properly designing a variable-stepsize integration scheme. Numerical experiments show that good results can be obtained by linking the integration stepsize to the total gravitational acceleration, with the stepsize increasing whenever the total gravitational acceleration decreases, and viceversa.

7.3. Remark. The computations reported here were done on a Unix Sun Workstation using the C++ programming language. In particular, the integrations were executed via a fifth-order Runge-Kutta-Fehlberg scheme.

8. Planetary and Mission Data

The gravitational constants for the Sun, Earth, and Mars are given by

$$\mu_S = 1.327\text{E}11, \quad \mu_E = 3.986\text{E}05, \quad \mu_M = 4.283\text{E}04 \text{ km}^3/\text{s}^2. \tag{31}$$

Earth and Mars travel around the Sun along orbits with average radii

$$r_E = 1.496\text{E}08, \quad r_M = 2.279\text{E}08 \text{ km}. \tag{32}$$

The associated average translational velocities and angular velocities (inertial coordinates) are given by

$$V_E = 29.78, \quad V_M = 24.13 \text{ km/s}, \tag{33a}$$

$$\omega_E = 0.9855, \quad \omega_M = 0.5241 \text{ deg/day}. \tag{33b}$$

In particular, the angular velocity difference between Earth and Mars is

$$\Delta\omega = \omega_E - \omega_M = 0.4614 \text{ deg/day}. \tag{34}$$

For the outgoing trip, the spacecraft is to be transferred from a low Earth orbit to a low Mars orbit; for the return trip, the spacecraft is to be transferred from a low Mars orbit to a low Earth orbit. The radii of the terminal orbits are

$$r_{LEO} = 6841, \qquad r_{LMO} = 3597 \text{ km}, \tag{35a}$$

corresponding to the altitudes

$$h_{LEO} = 463, \qquad h_{LMO} = 200 \text{ km}, \tag{35b}$$

since the Earth and Mars surface radii are given by

$$R_E = 6378, \qquad R_M = 3397 \text{ km}. \tag{35c}$$

The circular velocities at LEO and LMO (relative-to-planet coordinates) are given by

$$V_{LEO} = 7.633, \qquad V_{LMO} = 3.451 \text{ km/s}, \tag{36a}$$

and the corresponding escape velocities (relative-to-planet coordinates) are given by

$$\widetilde{V}_{LEO} = 10.795, \quad \widetilde{V}_{LMO} = 4.880 \text{ km/s}. \tag{36b}$$

9. Baseline Optimal Trajectory Results

In this section, we present the results obtained by solving the mathematical programming problems of Section 6 with the algorithm of Section 7 in light of the planetary and mission data of Section 8.

9.1. Outgoing Trip. The optimal LEO-to-LMO trajectory is shown in Figs. 2-3.

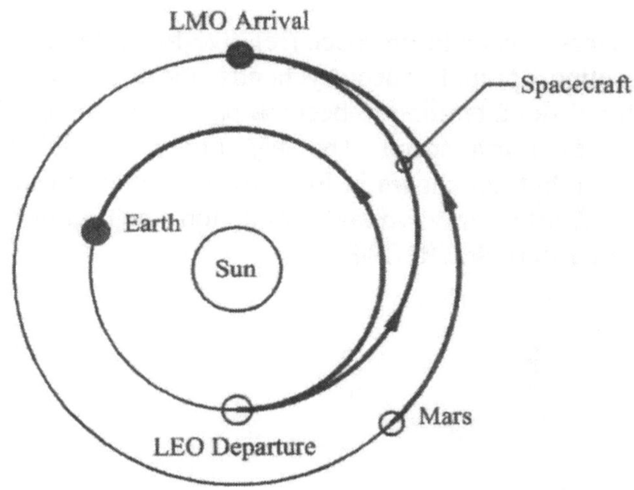

Fig. 2a. Optimal trajectory, outgoing trip, interplanetary space.

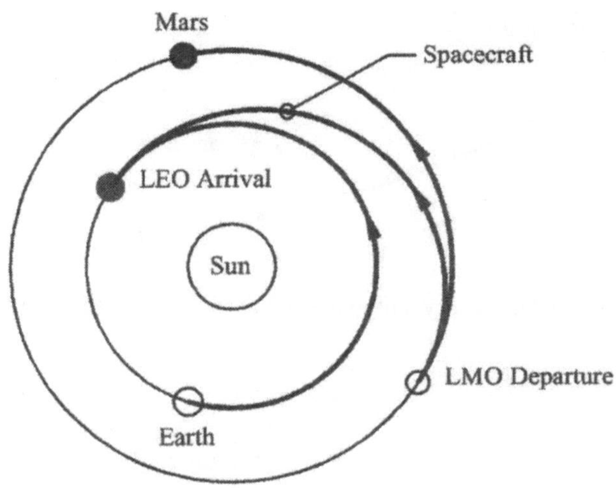

Fig. 2b. Optimal trajectory, return trip, interplanetary space.

Figure 2a refers to deep interplanetary space (Sun coordinates). The baseline optimal trajectory resembles a Hohmann transfer trajectory, but is not a Hohmann transfer trajectory, due to the disturbing influence of the gravitational fields of Earth and Mars on the terminal portions of the trajectory.

Figure 3a refers to near-Earth space (relative-to-Earth coordinates, first hour). The baseline optimal trajectory bends under the influence of the Earth gravitational field, tending to become parallel to the Earth trajectory at the end of near-Earth space. The asymptotic parallelism condition (hinted by Fig. 3a, but not shown in Fig. 3a) is reached toward the end of the first day (Earth gravitational attraction negligible w.r.t. Sun gravitational attraction). See Ref. 18.

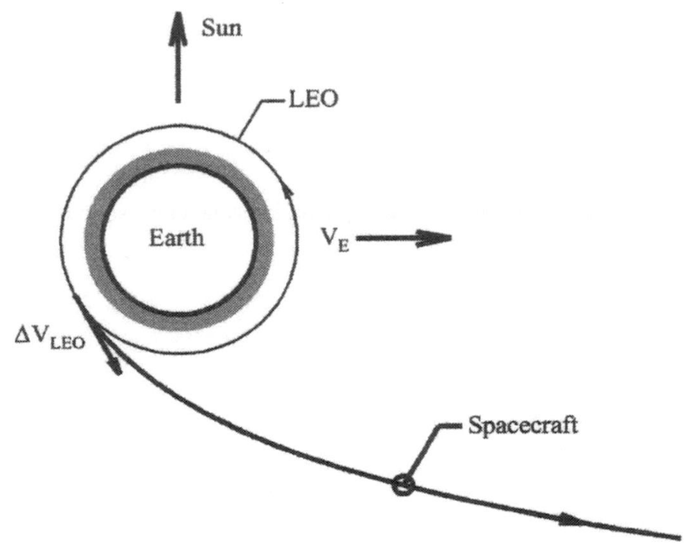

Fig. 3a. Optimal trajectory, outgoing trip, near-Earth space, first hour.

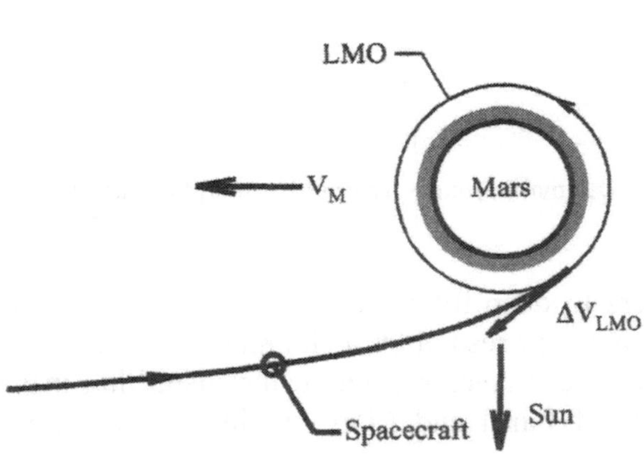

Fig. 3b. Optimal trajectory. outgoing trip, near-Mars space, last hour.

Figure 3b refers to near-Mars space (relative-to-Mars coordinates, last hour). In reverse time, the baseline optimal trajectory bends under the influence of the Mars gravitational field, tending to become parallel to the Mars trajectory at the beginning of near-Mars space. The asymptotic parallelism condition (hinted by Fig. 3b, but not shown in Fig. 3b) is reached at the beginning of the last day (Mars gravitational attraction negligible w.r.t. Sun gravitational attraction). See Ref. 18.

Major numerical results are given below:

(i) The terminal values of the Mars/Earth inertial phase angle difference are

$$\Delta\theta(0) = +43.9 \text{ deg}, \qquad \Delta\theta(\tau) = -75.1 \text{ deg}, \tag{37a}$$

meaning that Mars is ahead of Earth by nearly 44 deg at departure and behind Earth by nearly 75 deg at arrival.

(ii) The terminal values of the spacecraft/planet relative phase angle are

$$\phi_{PE}(0) = \theta_{PE}(0) - \theta_E(0) = -61.9 \text{ deg}, \tag{37b}$$

$$\phi_{PM}(\tau) = \theta_{PM}(\tau) - \theta_M(\tau) = -141.0 \text{ deg}, \tag{37c}$$

meaning that the accelerating velocity impulse at departure must be applied nearly 62 deg before the spacecraft becomes aligned with the Sun/Earth direction, while the decelerating velocity impulse at arrival must be applied nearly 141 deg before the spacecraft becomes aligned with the Sun/Mars direction.

(iii) The characteristic velocity components are

$$\Delta V_{LEO}(0) = 3.55 \text{ km/s}, \qquad \Delta V_{LMO}(\tau) = 2.10 \text{ km/s}, \tag{37d}$$

implying that the total characteristic velocity is

$$\Delta V = 5.65 \text{ km/s}. \tag{37e}$$

(iv) The terminal values of the spacecraft inertial phase angle are

$$\theta_P(0) = 0.0 \text{ deg}, \qquad \theta_P(\tau) = 179.0 \text{ deg}, \tag{37f}$$

implying that the angular travel of the spacecraft is

$$\theta_P(\tau) - \theta_P(0) = 179.0 \text{ deg}, \tag{37g}$$

which is within one degree of 180 deg, the value characterizing a Hohmann transfer trajectory.

(v) The transfer time is

$$\tau = 257.9 \text{ days} \cong 8.5 \text{ months} \cong 0.71 \text{ years}. \tag{37h}$$

9.2. Return Trip. The optimal LMO-to-LEO trajectory is shown in Figs. 2 and 4. Figure 2b refers to deep interplanetary space (Sun coordinates). The baseline optimal trajectory resembles a Hohmann transfer trajectory, but is not a Hohmann transfer trajectory, due to the disturbing influence of the gravitational fields of Mars and Earth on the terminal portions of the trajectory.

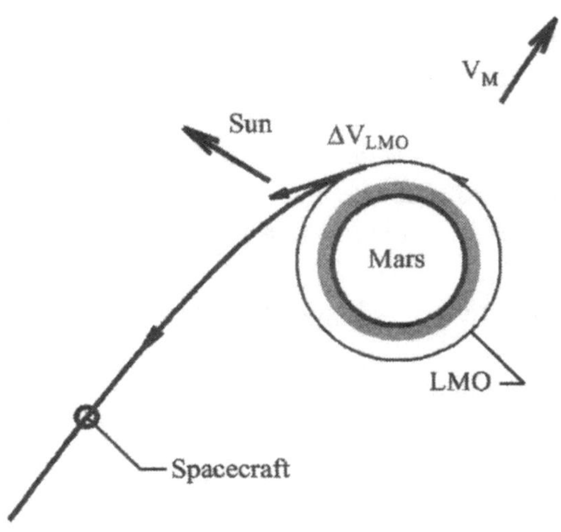

Fig. 4a. Optimal trajectory, return trip, near-Mars space, first hour.

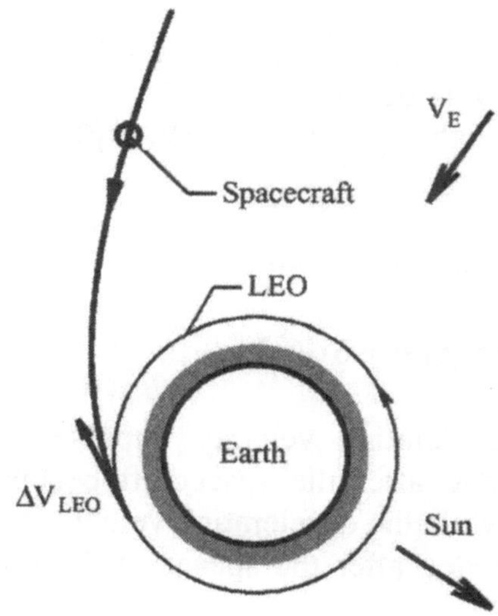

Fig. 4b. Optimal trajectory, return trip, near-Earth space, last hour.

Figure 4a refers to near-Mars space (relative-to-Mars coordinates, first hour). The baseline optimal trajectory bends under the influence of the Mars gravitational field, tending to become parallel to the Mars trajectory at the end of near-Mars space. The asymptotic parallelism condition (hinted by Fig. 4a, but not shown in Fig. 4a) is reached toward the end of the first day (Mars gravitational attraction negligible w.r.t. Sun gravitational attraction). See Ref. 18.

Figure 4b refers to near-Earth space (relative-to-Earth coordinates, last hour). In reverse time, the baseline optimal trajectory bends under the influence of the Earth gravitational field, tending to become parallel to the Earth trajectory at the beginning of near-Earth space. The asymptotic parallelism condition (hinted by Fig. 4b, but not shown in Fig. 4b) is reached at the beginning of the last day (Earth gravitational attraction negligible w.r.t. Sun gravitational attraction). See Ref. 18.

Major numerical results are given below:

(i) The terminal values of the Mars/Earth inertial phase angle difference are

$$\Delta\theta(0) = +\,75.0 \text{ deg}, \qquad \Delta\theta(\tau) = -\,43.9 \text{ deg}, \tag{38a}$$

meaning that Mars is ahead of Earth by nearly 75 deg at departure and behind Earth by nearly 44 deg at arrival.

(ii) The terminal values of the spacecraft/planet relative phase angle are

$$\phi_{PM}(0) = \theta_{PM}(0) - \theta_{M}(0) = +\,140.8 \text{ deg}, \tag{38b}$$

$$\phi_{PE}(\tau) = \theta_{PE}(\tau) - \theta_{E}(\tau) = +\,61.6 \text{ deg}, \tag{38c}$$

meaning that the accelerating velocity impulse at departure must be applied nearly 141 deg after the spacecraft becomes aligned with the Sun/Mars direction, while the decelerating velocity impulse at arrival must be applied nearly 62 deg after the spacecraft becomes aligned with the Sun/Earth direction.

(iii) The characteristic velocity components are

$$\Delta V_{LMO}(0) = 2.10 \text{ km/s}, \qquad \Delta V_{LEO}(\tau) = 3.55 \text{ km/s}, \tag{38d}$$

implying that the total characteristic velocity is

$$\Delta V = 5.65 \text{ km/s}. \tag{38e}$$

(iv) The terminal values of the spacecraft inertial phase angle are

$$\theta_{P}(0) = 0.0 \text{ deg}, \qquad \theta_{P}(\tau) = 179.1 \text{ deg}, \tag{38f}$$

implying that the angular travel of the spacecraft is

$$\theta_{P}(\tau) - \theta_{P}(0) = 179.1 \text{ deg}, \tag{38g}$$

which is within one degree of 180 deg, the value characterizing a Hohmann transfer trajectory.

(v) The transfer time is

$$\tau = 257.8 \text{ days} \cong 8.5 \text{ months} \cong 0.71 \text{ years}. \tag{38h}$$

9.3. Waiting Time. The waiting time in LMO is determined by the need to allow the Mars/Earth inertial phase angle difference to transition from the arrival value of the outgoing trip to the departure value of the return trip. In light of the previous results, the waiting time on Mars is

$$\tau_{\text{waiting}} = 454.3 \text{ days} = 14.9 \text{ months} = 1.24 \text{ years}. \tag{39}$$

Therefore, the total time for a round-trip LEO-to-LMO mission without delay time becomes

$$\tau_{*} = \tau_{\text{outgoing}} + \tau_{\text{waiting}} + \tau_{\text{return}}, \tag{40a}$$

and on account of the previous results,

$$\tau_{*} = 970 \text{ days} = 31.9 \text{ months} = 2.66 \text{ years}. \tag{40b}$$

9.4. Delay Time. If it is not possible to fire the rocket engines on the appropriate departure day for the return trip nor within the tolerance supplied by the so-called departure window (see Section 10), Eqs. (30) and (34) yield the delay time

$$\tau_{\text{delay}} = 779.2 \text{ days} = 25.6 \text{ months} = 2.13 \text{ years}. \tag{41}$$

Therefore, the total time for a round-trip LEO-to-LMO mission with delay time becomes

$$\tau_{**} = \tau_{\text{outgoing}} + \tau_{\text{waiting}} + \tau_{\text{delay}} + \tau_{\text{return}} \tag{42a}$$

$$= \tau_{*} + \tau_{\text{delay}},$$

and on account of (40b) and (41),

$$\tau_{**} = 1749.2 \text{ days} = 57.5 \text{ months} = 4.79 \text{ years}. \tag{42b}$$

9.5. Near-Mirror Property. In addition to the asymptotic parallelism property noted in Sections 9.1 and 9.2, the optimal trajectories of the outgoing and return trips have a near-mirror property, which emerges from the comparison of Eqs. (37a), (37b),(37c) with Eqs. (38a), (38b),(38c). These angular quantities can be grouped in pairs having nearly the same modulus but opposite sign for the outgoing and return trips. Also, the characteristic velocity components, total characteristic velocity, spacecraft angular travel, and transfer time are the same or nearly the same for the outgoing and return trips. The implication is that the optimal return trajectory can be obtained from the optimal outgoing trajectory via a sequential procedure of rotation, reflection, and inversion; see Ref. 19 for details. The near-mirror property extends to the restricted four-body problem the exact mirror property discovered by Miele for the restricted three-body problem in connection with the flight of a spacecraft in Earth-Moon space (Ref. 29).

10. Departure Windows for the Outgoing and Return Trips

In Section 6, we formulated the problems of minimizing the characteristic velocity for the outgoing trip (Problem P1) and return trip (Problem P2). In these problems, the vectors a, b, c appearing in Eqs. (25) and (28) have dimensions 7, 2, 5 respectively. The vector a includes the major parameters governing the transfer; the vector b includes the components of a that are fixed, namely, the radii of the terminal orbits; the vector c includes the components of a that must be optimized, namely, the terminal velocity impulses, transfer time, spacecraft/planet relative phase angle at departure, and Mars/Earth inertial phase angle difference at departure.

In this section, we modify the previous problems by assuming that the departure date η is fixed, hence by assuming that $\Delta\theta(0)$ is given. In the new problems, the vectors a, b, c have dimensions 7, 3, 4 respectively as can be seen by transferring $\Delta\theta(0)$ from (26c) to (26b) for the outgoing trip and from (28c) to (28b) for the return trip. Thus, one can formulate the following new problems:

Problem P3. For the outgoing trip, given the triplet (r_{LEO}, r_{LMO}, $\Delta\theta(0)$), minimize the performance index (25) w.r.t the parameters ΔV_{LEO}, ΔV_{LMO}, τ, $\theta_{PE}(0)$, subject to the constraints (14).

Problem P4. For the return trip, given the triplet (r_{LMO}, r_{LEO}, $\Delta\theta(0)$), minimize the performance index (27) w.r.t. the parameters ΔV_{LEO}, ΔV_{LMO}, τ, $\theta_{PM}(0)$, subject to the constraints (20).

While Problems P1 and P2 have two degrees of freedom, Problems P3 and P4 have one degree of freedom. For the outgoing trip, the true independent variable is the spacecraft/Earth relative phase angle at departure $\theta_{PE}(0)$; for the return trip, the true independent variable is the spacecraft/Mars relative phase angle at departure $\theta_{PM}(0)$.

The following relations connect the departure dates and the Mars/Earth phase angle differences:

$$[\Delta\theta(0)]_{P3} - [\Delta\theta(0)]_{P1} + (\omega_E - \omega_M)(\eta_{P3} - \eta_{P1}) = 0\,, \tag{43a}$$

$$[\Delta\theta(0)]_{P4} - [\Delta\theta(0)]_{P2} + (\omega_E - \omega_M)(\eta_{P4} - \eta_{P2}) = 0\,. \tag{43b}$$

Therefore, if one sets

$$\eta_{P1} = \eta_{P2} = 0\,, \qquad \eta_{P3} = \eta_{P4} = \eta\,, \tag{44}$$

and accounts for the baseline optimal trajectory results of Section 9, Eqs. (43) become

$$[\Delta\theta(0)]_{P3} = 43.9 - 0.4614\,\eta\,, \tag{45a}$$

$$[\Delta\theta(0)]_{P4} = 75.0 - 0.4614\,\eta\,, \tag{45b}$$

with η in days and $\Delta\theta(0)$ in degrees. Equations (45) establish a one-to-one correspondence between the departure date and the Mars/Earth inertial angle difference at departure.

By varying the departure date, hence by varying the Mars/Earth inertial phase angle difference at departure, one generates a one-parameter family of mathematical programming problems, whose solutions form the so-called departure windows for the outgoing and return trips. Note that, if $\eta = 0$ in Eq. (45a), the solution of Problem P3 reduces to that of Problem P1; also note that, if $\eta = 0$ in Eq. (45b), the solution of Problem P4 reduces to that of Problem P2.

10.1. Results. For the outgoing and return trips, Tables 1 and 2 list the departure date, Mars/Earth inertial phase angle difference at departure, spacecraft/planet relative phase angle at departure, spacecraft angular travel, characteristic velocity components at departure and arrival, total characteristic velocity, flight time, and Mars/Earth inertial phase angle difference at arrival. In these tables, the central column refers to the baseline optimal trajectory; the left column refers to the suboptimal trajectory generated via anticipated departure by nearly 6 weeks; the right column refers to the suboptimal trajectory generated via delayed departure by nearly 3 weeks.

Major comments are as follows. For the suboptimal trajectories, the Mars/Earth inertial phase angle difference at departure increases with early departure and decreases with late departure; the angular travel and flight time increase with early departure and decrease with late departure; the characteristic velocity components and total characteristic velocity increase with both early and late departures. The above statements hold for both the outgoing and return trips. Finally, it must be noted that, for the suboptimal trajectories of both the outgoing and return trips, the asymptotic parallelism property no longer holds for the departure branch, but still holds for the arrival branch. On the other hand, the near-mirror property no longer holds.

Table 1. Outgoing trip: Departure window results.

η	– 43.5	0.0	+ 21.3	day
$\Delta\theta(0)$	64.0	43.9	34.0	deg
$\phi_{PE}(0)$	0.4	– 61.9	– 91.6	deg
$\theta_P(\tau) - \theta_P(0)$	209.4	179.0	163.0	deg
ΔV_{LEO}	3.95	3.55	3.70	km/s
ΔV_{LMO}	2.20	2.10	2.14	km/s
ΔV	6.15	5.65	5.84	km/s
τ	278.3	257.9	246.1	day
$\Delta\theta(\tau)$	– 64.4	–75.1	–79.6	deg

11. Comments and Conclusions

From the previous analysis, the following comments and conclusions emerge.

(i) The Mars mission is difficult because of the large distances involved. In a round-trip LEO-LMO-LEO mission, the curvilinear distance traveled along the trajectory exceeds one billion kilometers. At some point of the trajectory, the spacecraft/Earth distance becomes larger than the Earth/Sun distance.

(ii) The extremely long journey requires a long flight time, namely, 0.71 years for the outgoing trip, 0.71 years for the return trip, 1.24 years waiting in LMO, plus a delay time of 2.13 years if the spacecraft is unable to fire the rocket engines within the departure window tolerance for return. The total round-trip time is 2.66 years without time delay and 4.79 years with time delay.

(iii) If one converts the characteristic velocity results into mass ratios

Table 2. Return trip: Departure window results.

η	− 43.3	0.0	+ 21.6	day
$\Delta\theta(0)$	95.0	75.0	65.0	deg
$\phi_{PM}(0)$	175.8	140.8	123.1	deg
$\theta_P(\tau) - \theta_P(0)$	193.1	179.1	171.6	deg
ΔV_{LMO}	2.29	2.10	2.17	km/s
ΔV_{LEO}	3.57	3.55	3.56	km/s
ΔV	5.86	5.65	5.73	km/s
τ	292.3	257.8	240.1	day
$\Delta\theta(\tau)$	− 39.9	− 43.9	− 45.7	deg

using typical values of the spacecraft structural factor and engine specific impulse, it can be seen that the required mass ratio for a round-trip LEO-LMO-LEO mission is about 20. This means that, to return the mass of 1 kg to LEO, we need the mass of 20 kg at the departure from LEO.

If one includes the ascent from the Earth surface to LEO, the required mass ratio becomes of order 300. If one further includes the ascent from the Mars surface to LMO, the required mass ratio becomes of order 1000. This means that, to return the mass of 1 kg to Earth, we need the mass of 1000 kg at the departure from Earth.

(iv) With reference to (iii), the required mass ratios can be decreased via the use of aeroassisted orbital transfer maneuvers, also called aerobraking maneuvers. See Refs. 30-33 for recent work on these special maneuvers.

(v) The best trajectory is the baseline optimal trajectory. For the outgoing trip, Mars must be ahead of Earth by nearly 44 deg at departure and the accelerating velocity impulse must be applied 62 deg before the spacecraft become aligned with the Sun/Earth direction. For the return trip, Mars must be ahead of Earth by nearly 75 deg at departure and the accelerating velocity impulse must be applied 141 deg after the spacecraft becomes aligned with the Sun/Mars direction.

(vi) The baseline optimal trajectory resembles a Hohmann transfer trajectory, but is not a Hohmann transfer trajectory, owing to the disturbing influence exerted by the gravity fields of Earth and Mars on the terminal branches of the trajectory.

(vii) An important property of the baseline optimal trajectory is the asymptotic parallelism property: For optimal transfer, the spacecraft inertial velocity must be parallel to the inertial velocity of the closest planet (Earth or Mars) at the entrance to and exit from deep interplanetary space. This asymptotic parallelism occurs at the end of the first day and at the beginning of the last day for both the outgoing and return trips.

(viii) Another property of the baseline optimal trajectory is the near-mirror property. The return trajectory can be obtained from the outgoing trajectory via a sequential procedure of rotation, reflection, and inversion. This property extends to the restricted four-body problem the exact mirror property found for the restricted three-body problem in connection with flight of a spacecraft in Earth/Moon space (Ref. 29).

(ix) Departure window trajectories are next-to-best trajectories. They are suboptimal trajectories obtained by changing the departure date, hence changing the Mars/Earth inertial phase angle difference at departure. For the departure window trajectories, anticipated departure yields longer flight time and wider angular travel; delayed departure yields shorter flight time and narrower angular travel.

(x) For the departure window trajectories, the asymptotic parallelism property no longer holds in the departure branch, but still holds in the arrival branch. On the other hand, the near-mirror property no longer holds.

(xi) While the present analysis is valid for both robotic and manned missions, this author believes that, on account of the extremely long flight times [see (ii)], robotic missions should be preferred for the time being. Manned missions are extremely difficult and should not be attempted unless one solves first all the problems that need to be solved to ensure the survival of the astronauts in space and time.

(xii) It must be emphasized that the present study is preliminary. Additional studies are under way to account for the ellipticity of the motion of Earth and Mars around Sun. Further studies are under way to account for the fact that the Earth and Mars orbital planes are not identical. Also, aerobraking maneuvers are being considered as a means to reduce propellant consumption through penetration of the Mars atmosphere in the outgoing trip and penetration of the Earth atmosphere in the return trip (Refs. 30-33).

References

1. LINDORFER, W., and MOYER, H. G., *Application of a Low Thrust Trajectory Optimization Scheme to Planar Earth-Mars Transfer*, ARS Journal, Vol. 32, pp. 260-262, 1962.

2. ANONYMOUS, *The Viking Mission to Mars,* Martin Marietta Corporation, Denver, Colorado, 1975.

3. LECOMPTE, M., *New Approaches to Space Exploration*, The Case for Mars, Edited by P. J. Boston, Univelt, San Diego, California, pp. 35-37, 1984.

4. NIEHOFF, J. C., *Pathways to Mars: New Trajectory Opportunities,* NASA Mars Conference, Edited by D. B. Reiber, Univelt, San Diego, California, pp. 381-401, 1988.

5. ROY, A. E., *Orbital Motion*, Adam Hilger, Bristol, England, 1988.

6. STRIEPE, S. A., BRAUN, R. D., POWELL, R. W., and FOWLER, W. T., *Influence of Interplanetary Trajectory Selection on Mars Atmospheric Entry Velocity,* Journal of Spacecraft and Rockets, Vol. 30, No. 4, pp. 426-430, 1993.

7. TAUBER, M., HENLINE, W., CHARGIN, M., PAPADOPOULOS, P., CHEN, Y., YANG, L., and HAMM, K., *Mars Environmental Survey Probe, Aerobrake Preliminary Design Study*, Journal of Spacecraft and Rockets, Vol. 30, No.4, pp. 431-437, 1993.

8. BRAUN, R. D., POWELL, R. W., ENGELUND, W. C., GNOFFO, P. A., WEILMUENSTER, K. J., and MITCHELTREE, R. A., *Mars Pathfinder Six-Degree-of-Freedom Entry Analysis*, Journal of Spacecraft and Rockets, Vol. 32, No.6, pp. 993-1000, 1995.

9. GURZADYAN, G. A., *Theory of Interplanetary Flights*, Gordon and Breach Publishers, Amsterdam, Netherlands, 1996.

10. LEE, W., and SIDNEY, W., *Mission Plan for Mars Global Surveyor*, Spaceflight Mechanics 1996, Edited by G. E. Powell, R. H. Bishop, J. B. Lundberg, and R. H. Smith, Univelt, San Diego, California, pp. 839-858, 1996.

11. SPENCER, D. A., and BRAUN, R. D., *Mars Pathfinder Atmospheric Entry: Trajectory Design and Dispersion Analysis*, Journal of Spacecraft and Rockets, Vol. 33, No. 5, pp. 670-676, 1996.

12. STRIEPE, S. A., and DESAI, P. N., *Piloted Mars Missions Using Cryogenic and Storable Propellants*, Journal of the Astronautical Sciences, Vol. 44, No. 2, pp.207-222, 1996.

13. WAGNER, L. A., Jr., and MUNK, M. M., *MISR Interplanetary Trajectory Design*, Spaceflight Mechanics 1996, Edited by G. E. Powell, R. H. Bishop, J. B. Lundberg, and R. H. Smith, Univelt, San Diego, California, pp. 859-876, 1996.

14. WERCINSKI, P. F., *Mars Sample Return: A Direct and Minimum-Risk Design*, Journal of Spacecraft and Rockets, Vol. 33, No.3, pp. 381-385, 1996.

15. LOHAR, F. A., MISRA, A. K., and MATEESCU, J. D., *Mars-Jupiter Aerogravity Assist Trajectories for High-Energy Missions*, Journal of Spacecraft and Rockets, Vol. 34, No. 1, pp. 16-21, 1997.

16. MIELE, A., and WANG, T., *Optimal Trajectories for Earth-Mars Flight,* Journal of Optimization Theory and Applications, Vol. 95, No. 3, pp. 467-499, 1997.

17. MIELE, A., and WANG, T., *Optimal Transfers from an Earth Orbit to a Mars Orbit*, Acta Astronautica, Vol. 45, No. 3, pp.119-133, 1999.

18. MIELE, A., and WANG, T., *Optimal Trajectories and Asymptotic Parallelism Property for Round-Trip Mars Missions*, Proceedings of the 2nd International Conference on Nonlinear Problems in Aviation and Aerospace, Edited by S. Sivasundaram, European Conference Publications, Cambridge, England, Vol. 2, pp. 507-539, 1999.

19. MIELE, A., and WANG, T., *Optimal Trajectories and Mirror Properties for Round-Trip Mars Missions*, Acta Astronautica, Vol. 45, No. 11, pp. 655-668, 1999.

20. MIELE, A., HUANG, H.Y., and HEIDEMAN, J. C., *Sequential Gradient-Restoration Algorithm for the Minimization of Constrained Functions: Ordinary and Conjugate Gradient Versions*, Journal of Optimization Theory and Applications, Vol. 4, No. 4, pp. 213-243, 1969.

21. MIELE, A., TIETZE, J. L., and LEVY, A. V., *Comparison of Several Gradient Algorithms for Mathematical Programming Problems*, Omaggio a Carlo Ferrari, Edited by G. Jarre, Libreria Editrice Universitaria Levrotto e Bella, Torino, Italy, pp. 521-536, 1974.

22. MIELE, A., PRITCHARD, R. E., and DAMOULAKIS, J. N., *Sequential Gradient-Restoration Algorithm for Optimal Control Problems*, Journal of Optimization Theory and Applications, Vol. 5, No.4, pp. 235-282, 1970.

23. MIELE, A., TIETZE, J. L, and LEVY, A. V., *Summary and Comparison of Gradient-Restoration Algorithms for Optimal Control Problems*, Journal of Optimization Theory and Applications, Vol. 10, No. 6, pp. 381-403, 1972.

24. MIELE, A., and DAMOULAKIS, J. N., *Modifications and Extensions of the Sequential Gradient-Restoration Algorithm for Optimal Control Theory*, Journal of the Franklin Institute, Vol. 294, No. 1, pp. 23-42, 1972.

25. MIELE, A., and WANG, T, *Primal-Dual Properties of Sequential Gradient-Restoration Algorithms for Optimal Control Problems, Part 1: Basic Problem*, Integral Methods in Science and Engineering, Edited by F. R. Payne et al, Hemisphere Publishing Corporation, Washington, DC, pp. 577-607, 1986.

26. MIELE, A., and WANG, T., *Primal-Dual Properties of Sequential Gradient-Restoration Algorithms for Optimal Control Problems, Part 2: General Problem*, Journal of Mathematical Analysis and Applications, Vol. 119, Nos. 1-2, pp. 21-54, 1986.

27. MIELE, A., WANG, T., and BASAPUR, V. K., *Primal and Dual Formulations of Sequential Gradient-Restoration Algorithms for Trajectory Optimization Problems*, Acta Astronautica, Vol. 13, No. 8, pp. 491-505, 1986.

28. RISHIKOF, B. H., MCCORMICK, B. R, PRITCHARD, R. E., and SPONAUGLE, S. J., *SEGRAM: A Practical and Versatile Tool for Spacecraft Trajectory Optimization*, Acta Astronautica, Vol. 26, Nos. 8-10, pp. 599-609, 1992.

29. MIELE, A., *Theorem of Image Trajectories in the Earth-Moon Space*, Astronautica Acta, Vol.4, No. 5, pp. 225-232, 1960.

30. MIELE, A., and WANG, T., *Nominal Trajectories for the Aeroassisted Flight Experiment*, Journal of the Astronautical Sciences, Vol. 41, No.2, pp. 139-163, 1993.

31. MIELE, A., *Recent Advances in the Optimization and Guidance of Aeroassisted Orbital Transfers*, The 1st John V. Breakwell Memorial Lecture, Acta Astronautica, Vol. 38, No. 10, pp. 747-768, 1996.

32. MIELE, A., and WANG, T., *Robust Predictor-Corrector Guidance for Aeroassisted Orbital Transfer*, Journal of Guidance, Control, and Dynamics, Vol. 19, No. 5, pp. 1134-1141, 1996.

33. MIELE, A., and WANG, T., *Near-Optimal Highly Robust Guidance for Aeroassisted Orbital Transfer*, Journal of Guidance, Control, and Dynamics, Vol. 19, No.3, pp. 549-556, 1996.

4

Design and Test of an Experimental Guidance System with a Perspective Flight Path Display

G. SACHS[1]

Abstract. Design issues of a guidance system displaying visual information in a 3-dimensional format to the pilot for improving manual flight path control are considered. A basic concept of such a synthetic vision system is described, yielding an integrated presentation of the command flight path and the terrain, supplemented by other guidance elements. The imagery is generated by a computer in real time with an adequate update rate, using attitude and position data from a precision navigation system. This basic synthetic vision system was flight tested in an experimental program consisting of several test series, with demanding flight tasks aiming at different control aspects. The flight test results show that the synthetic vision system enabled the pilot to control precisely the aircraft and hold it on the command trajectory. Furthermore, an extended 3-dimensional guidance display concept is considered which employs a predictor indicating the future position of the aircraft at a specified time ahead. Design issues are described for achieving a predictor aircraft system requiring minimum pilot compensation. Results from pilot-in-the-loop simulation experiments are presented which provide a verification of the design considerations.

Professor and Director, Institute of Flight Mechanics and Flight Control, Technische Universität München, 85747 Garching, Germany.

Key Words. Perspective flight path display, synthetic vision, flight path predictor, manual flight path control, aircraft guidance.

Nomenclature

e = error,
g = acceleration of gravity,
K = gain,
L_{δ_a} = roll moment due to roll control input,
s = Laplace operator,
T = time constant,
$Y(s)$ = transfer function,
y = lateral coordinate,
Δy = perturbation of y,
δ_a = roll control,
ζ = damping ratio,
τ_e = effective time delay,
ϕ = roll angle,
χ = azimuth angle,
ω = frequency.

1. Introduction

Innovative approaches for the cockpit instrumentation of aircraft are displays which present guidance information in a 3-dimensional format to the pilot. They show the future flight path in a perspective form and may additionally depict a terrain imagery. Such displays, which are known as tunnel or highway-in-the-sky displays, offer a fundamental enhancement in the visual information of the pilot because they provide status and command information not only of actual but also of future flight situations. Furthermore, perspective flight path displays present the information in a descriptive format and allow holistic perception. The visual information can be perceived intuitively and directly by the pilot and the scanning workload decreased. As a result, the mental effort for reconstructing the spatial and temporal situation may be reduced substantially when compared with current instrumentation.

Results from recent research including theoretical investigations as well as simulation experiments and flight tests show that significant improvements in aircraft guidance and control can be achieved with displays presenting the flight path and other relevant information in a 3-dimensional format (Refs. 1-18). The flight test verification includes the worldwide first landing of an aircraft with a pictorial display presenting 3-dimensional guidance information (synthetic vision) as the only visual information for the pilot (Refs. 14, 15).

It is the purpose of this paper to describe design issues of perspective flight path displays and to present experimental results from simulation and flight tests.

2. Basic Concept of Three-Dimensional Guidance Display

The basic concept of the 3-dimensional guidance display comprises an integrated presentation of the flight path and the terrain, supplemented by other guidance elements. Such a display featuring synthetic vision includes the following constituents (Fig. 1): 3-dimensional guidance information; pictorial presentation of outside world; precision navigation.

Central element of the 3-dimensional guidance information is the perspective flight path presentation in the form of a tunnel (Fig. 2). Further guidance elements of primary significance are displayed in an integrated manner. Indication of the command flight path provides the pilot with a preview of the future trajectory. With command information and preview available, the pilot can use this preview to structure a control feedforward. This is illustrated in Fig. 3, which shows a simplified model for describing general pathways of the human controller operating on visually sensed inputs and exerting manual control outputs. Different control modes are possible, one of which is compensatory control applied as a closed-loop control for regulation tasks. The other control mode is pursuit/preview control which is possible because of command information and preview.

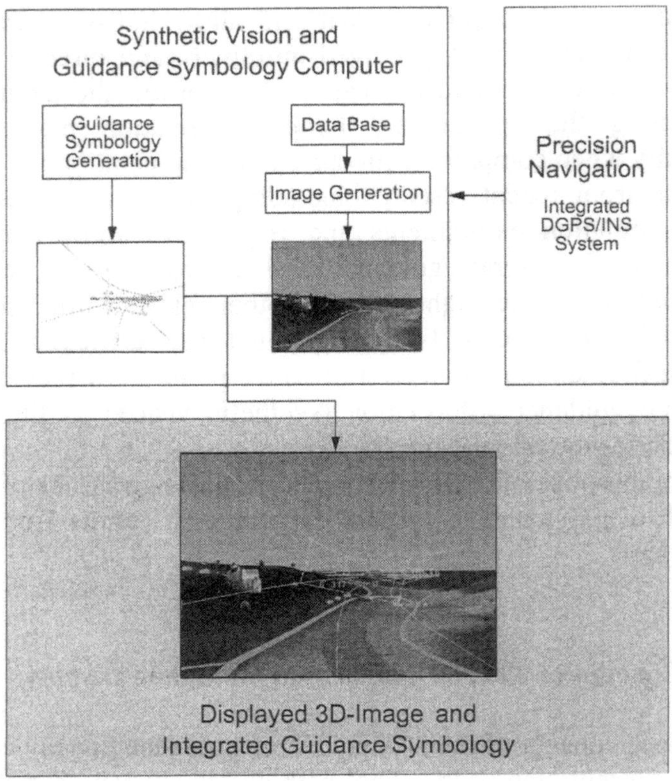

Fig. 1. Concept of 3-dimensional guidance display.

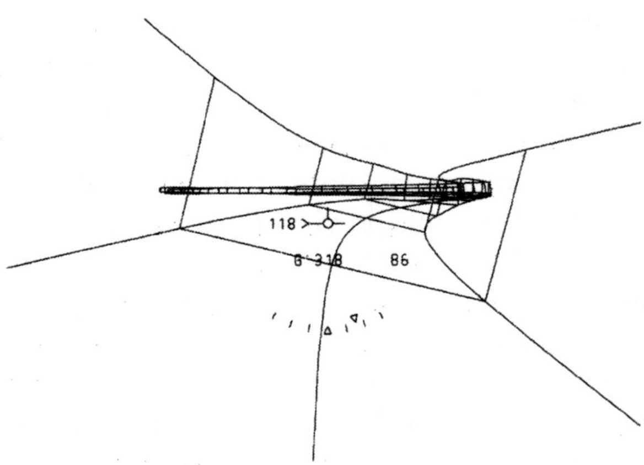

Fig. 2. Presentation of command flight path (tunnel) and other guidance elements.

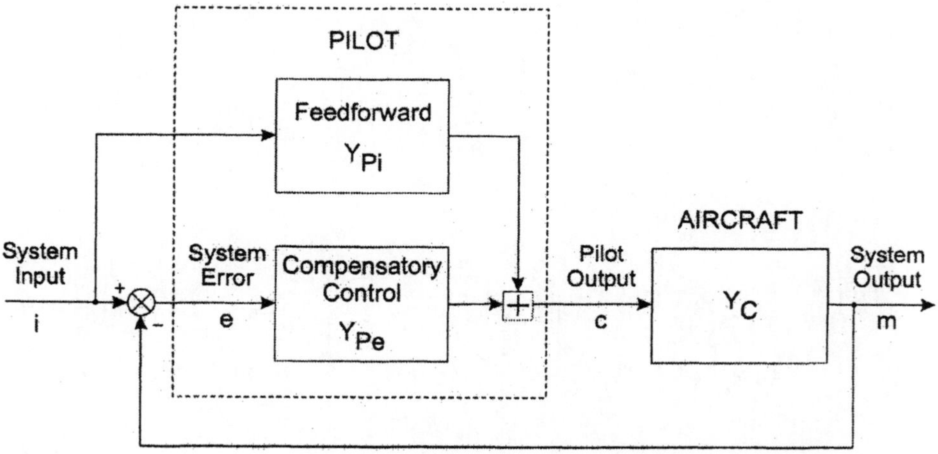

Fig. 3. Model for pilot-aircraft system.

The pictorial presentation of the outside world comprises an image of the terrain, including all relevant information about its elevation and features. This is illustrated in Fig. 4, which shows the integrated presentation of the outside world image and the guidance tunnel. Two groups of data are used for generating the outside world image: terrain elevation and feature analysis data. The terrain elevation data are referenced to a grid structure the elements of which have a size of 3" × 3" or 1" × 1" (Fig. 5). A grid element represents an area of about 90 m × 60 m (or 30 m × 20 m) at the geographical latitude of the areas where the flight test took place. Three data groups are applied for describing the terrain features (Fig. 6): point features (buildings, bridges, power line pylons, etc.); linear features (roads, railways, rivers, etc.); areal features (cities, forests, lakes, etc.).

A special treatment of terrain elevation and features is applied for areas where the aircraft operates close to the ground, like airports. It yields a precise modeling as regards location, elevation, dimensions, objects, etc.

The precision navigation system provides the synthetic vision computer with position and attitude data (Fig. 1). This is necessary for generating an image according to the actual field of view of the pilot. In the flight tests, a high-precision navigation system was applied using

Fig. 4. Integrated presentation of terrain and command flight path (tunnel).

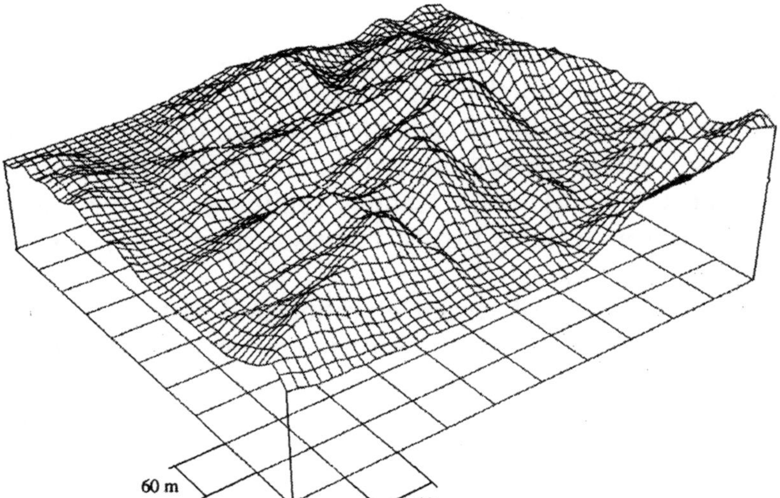

Fig. 5. Modelling of terrain elevation.

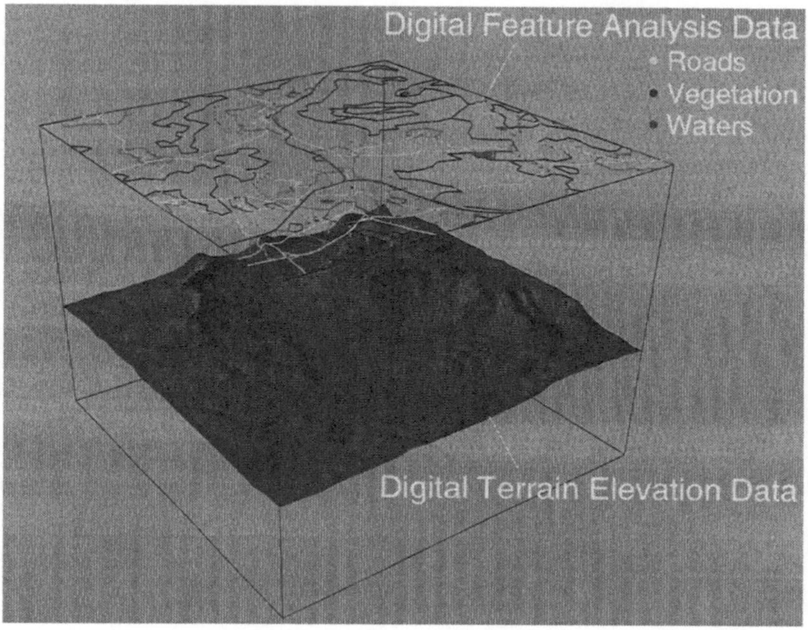

Fig. 6. Modelling of outside world image using terrain elevation and feature data.

differential global positioning and inertial sensor navigation data. The navigation system was operated in local and wide area DGPS modes for transmitting the differential correction data (Fig. 7). The local area DGPS mode was used in flight tests for terminal flight operations (approach and landing) using a customized ultra high frequency data link. The GPS ground reference station was located close to the runway. The wide area DGPS mode was applied in nonterminal flight tests (flights in river valleys and mountainous areas) using a low-frequency transmitting technique for providing the correction signal. Because of the low-frequency transmitting technique, it was possible to receive the correction data without having to cope with hiding effects due to terrain formations. The technique was developed by the Institute of Applied Geodesy in Potsdam, Germany (Ref. 19).

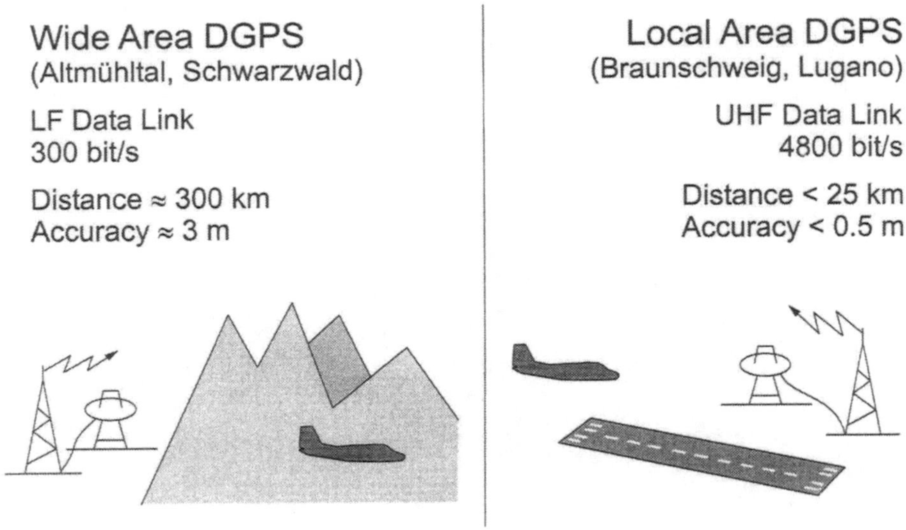

Fig. 7. Wide and local area DGPS modes.

3. Flight Test Results for Basic Three-Dimensional Guidance Display

A series of flight tests was performed aiming at a wide range of guidance applications of the 3D-guidance display system, featuring the above guidance information and terrain imagery. Demanding control tasks were specified and investigated. The test program consisted of five flight test series:

(i) precision approach and landing flight tests, Braunschweig Airport, Germany, October 10-14, 1994;

(ii) low-level flight tests in a highly curved, narrow river valley, Altmühl river, Germany, December 12-16, 1994;

(iii) curved and steep approaches in mountainous area, Lugano airport, Switzerland, July 31 - August 4, 1995;

(iv) curved/steep/short approaches and low-level and terrain-following flights in a mountainous area, Offenburg/Schwarzwald, Germany, March 18-22, 1996;

(v) curved/steep approaches and curved trajectory-following flights in a mountainous area, Freiburg/Schwarzwald, Germany, July 7-10, 1997.

An overview of the flight test areas is provided by Fig. 8, which shows the locations in Germany and Switzerland. The place of the ground reference station for the wide area DGPS mode is also depicted.

The vehicle which is used in the flight test program is a twin engine Dornier 128; it is operated by the Institute of Flight Guidance and Control of the Technische Universität Braunschweig as a research aircraft (Fig. 9). The aircraft is equipped with a high precision navigation system which was developed by this Institute. The high navigation performance is achieved by coupling differential global positioning and inertial sensor systems to yield an integrated precision navigation system (Ref. 20). In addition, computer and filter algorithms including error modeling are applied. Thus, it is possible to achieve a high precision for static as well as dynamic behavior.

Fig. 8. Flight test areas in Germany and Switzerland and ground reference station for wide area DGPS.

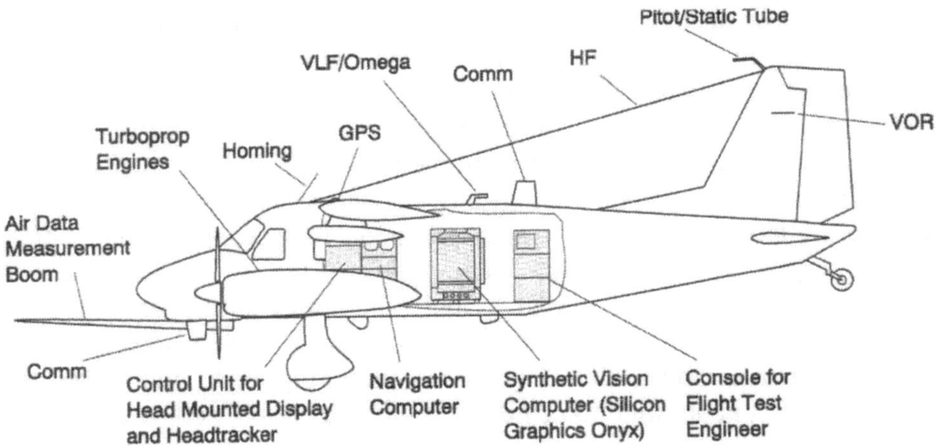

Fig. 9. Aircraft with equipment used in flight tests.

Results representative for the flight tests are presented in the following. They are from the low-level flight tests in the Altmühl river valley. The test course depicted in Fig. 10 shows that the Altmühl river area represents a demanding test environment because of the highly curved and narrow river valley with steep banks. The control task was to follow precisely the command trajectory, which was referenced to the course of the river, at an height of 100 m above the river as authorized by the flight safety agency.

The flight test results presented in Figs. 11 and 12 show that the pilot followed precisely the command trajectory, with only small deviations in both the vertical and lateral directions. This holds generally for the whole of the flight test course of about 70 km, and particularly for those sections where the control tasks were very demanding in the vertical or lateral direction. The command trajectory was indicated in the 3-dimensional guidance display by a tunnel image, as shown in Fig. 4 for a flight condition of the tests in the Altmühl river valley. From the results presented in Figs. 11 and 12, it follows that the aircraft stayed well within the tunnel.

For the motion in the vertical direction, there are three sections of particular interest because evasive maneuvers were necessary (Fig. 11). In two sections, electrical power lines intersect the river valley. This was shown in the 3-dimensional guidance display, with a corresponding change in the course of the tunnel. In a third section, there is a river bend so tight that it could not be followed by the aircraft. For this part of the trajectory, an evasive maneuver was specified according to which the pilot left the river valley, flew over the bank at the riverside and entered again the river valley afterward.

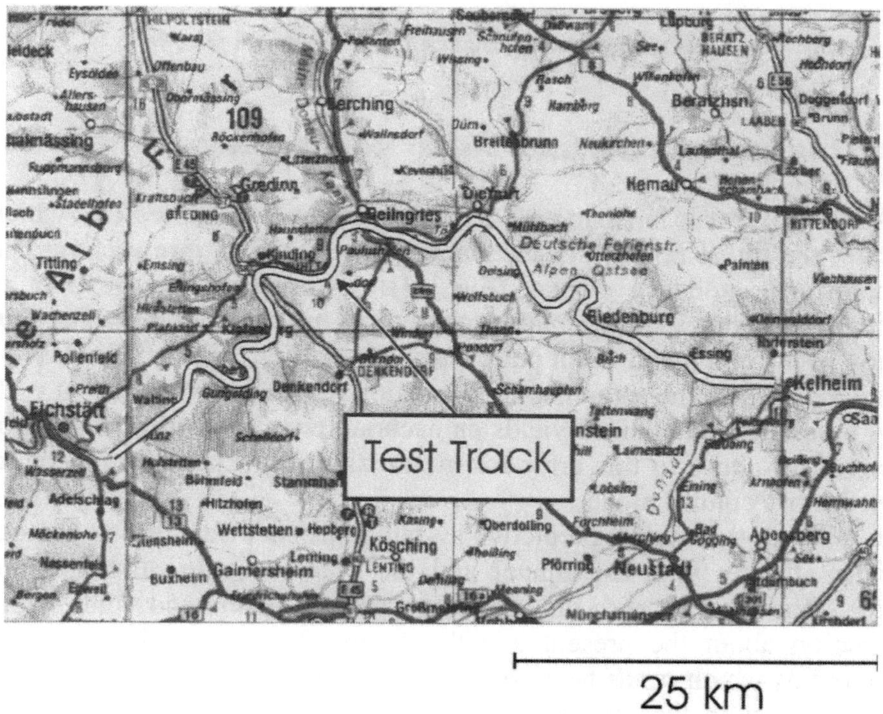

Fig. 10. Flight test area in Altmühl river valley.

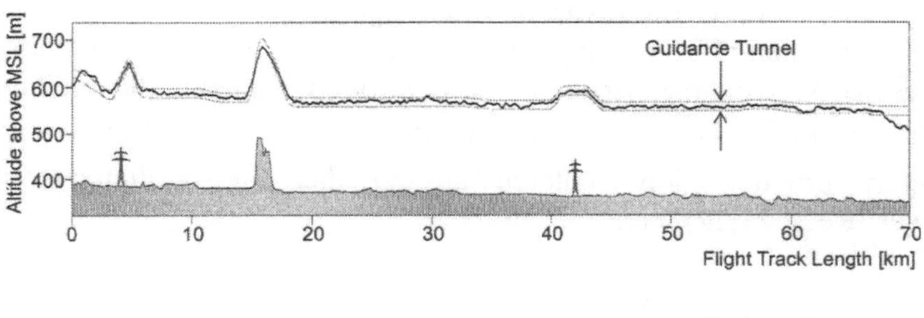

Fig. 11. Vertical path control during test flight in Altmühl river valley.

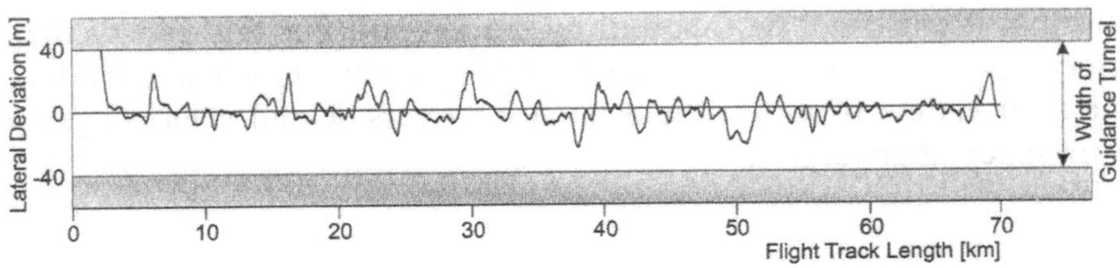

Fig. 12. Lateral path control during test flight in Altmühl river valley.

4. Three-Dimensional Guidance Display with Predictor

An improvement in flight path control is possible by a predictor which indicates the future position of the aircraft at a specified time ahead in the 3-dimensional guidance display (Fig. 13). This is because the pilot is provided with precise information about the future aircraft position in relation to the command flight path. As shown in Fig. 13, the deviation of the predictor from the reference cross section of the command flight path at the prediction time ahead yields an accurate error indication. The pilot can act in response to this error for minimizing flight path deviations in compensatory control mode.

In general, the overall predictive system consists of the 3-dimensional guidance display with the tunnel and the predictor, the pilot and the aircraft. The tunnel and the predictor present command and status information about the present and the future. There are pilot-centered requirements which result from the presence of the human operator in the control loop, with the objective to achieve an overall predictive system requiring minimum pilot compensation.

For achieving this objective, the predictive system should be constructed to require no low-frequency lead equalization for the pilot and to permit pilot-loop closure over a wide range of gains. This requirement can be met when the equalizations and gains are selected so that the effective transfer characteristic of the controlled element, the predictor-aircraft system $Y_{PR}(s)Y_C(s)$, approximates a pure integration over an adequately broad region centered around the pilot-predictor-aircraft crossover (Refs. 21,22), i.e.,

$$Y_{PR}(s)Y_C(s) = \frac{K}{s}. \tag{1}$$

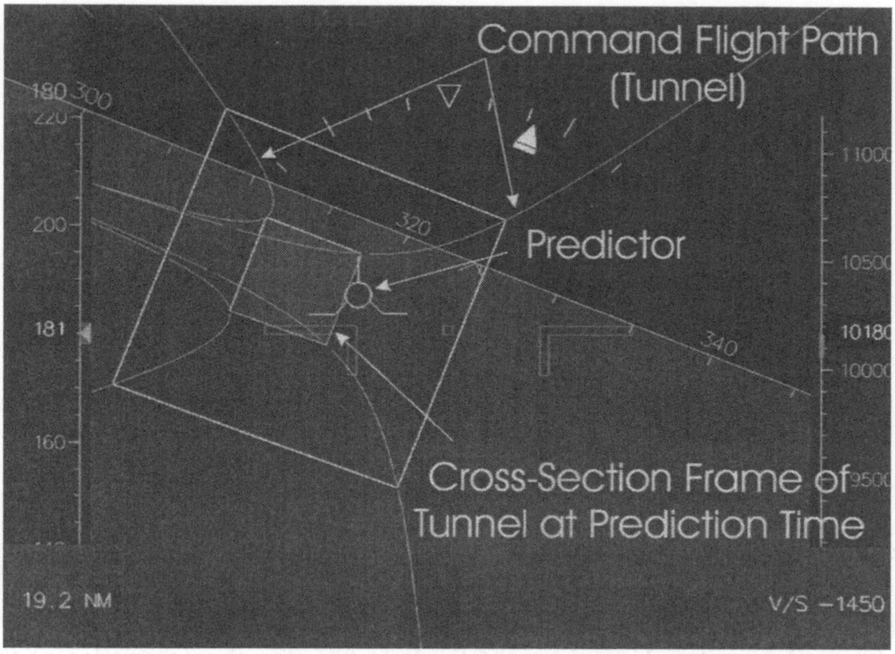

Fig. 13. Perspective flight path display with predictor.

This relation describes the desired dynamic characteristics of the predictor-aircraft system as the controlled element. It can be used as key requirement for designing the predictor to achieve appropriate dynamic characteristics of the closed-loop pilot-predictor-aircraft system.

Besides this manual control-related predictor issue, there is another point which is concerned with the role of the predictor as an indicator of the future aircraft position. A realistic indication of the future aircraft position can be considered a requirement for face validity according to which the status information presented by the predictor in the 3-dimensional guidance display should correspond to the actual situation. Thus, geometric and kinematic relations come into consideration for describing the continuation of the flight path to which the predicted position can be referenced. This is illustrated in Fig. 14, which shows a model for describing the continuation of the flight path in the lateral direction, with particular reference to the situation at the prediction time ahead.

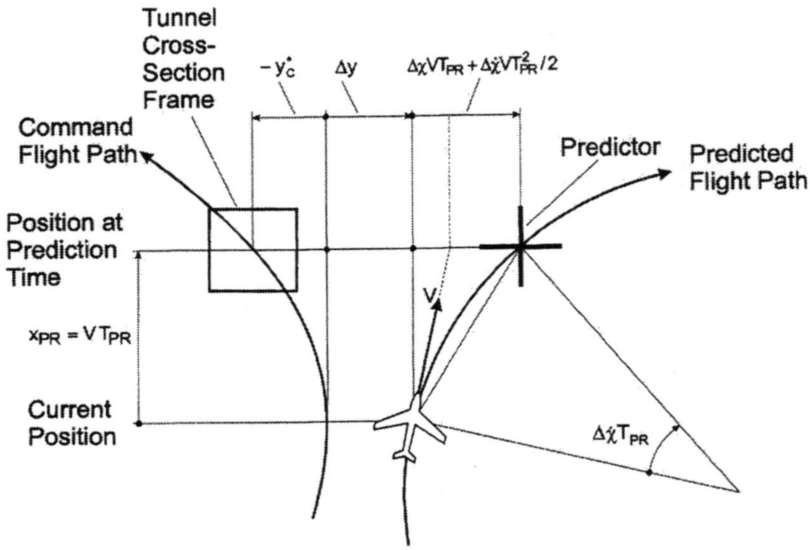

Fig. 14. Lateral displacement of predictor position (circular flight path continuation).

The pilot-centered requirement for best transfer characteristics of the predictor-aircraft system, supplemented by the face validity considerations, forms the basis for the predictor control law. With reference to the block diagram in Fig. 15, the predictor law for lateral flight path control can be constructed to yield

$$Y_{PR} = \frac{e_{PR}(s)}{\mathrm{D}f\,(s)} = K_{PR}\frac{\left(K_{\dot{\phi}}s^3 + g(T_{PR}^2/2)s^2 + gT_{PR}s + 1\right)}{s^2}, \qquad (2)$$

where T_{PR} is the prediction time related to the predictor position. Selecting for the roll rate gain

$$K_{\dot{\phi}} \approx \frac{g}{2}T_R T_{PR}^2, \qquad (3)$$

and applying the aircraft dynamics model valid for the frequency region of concern

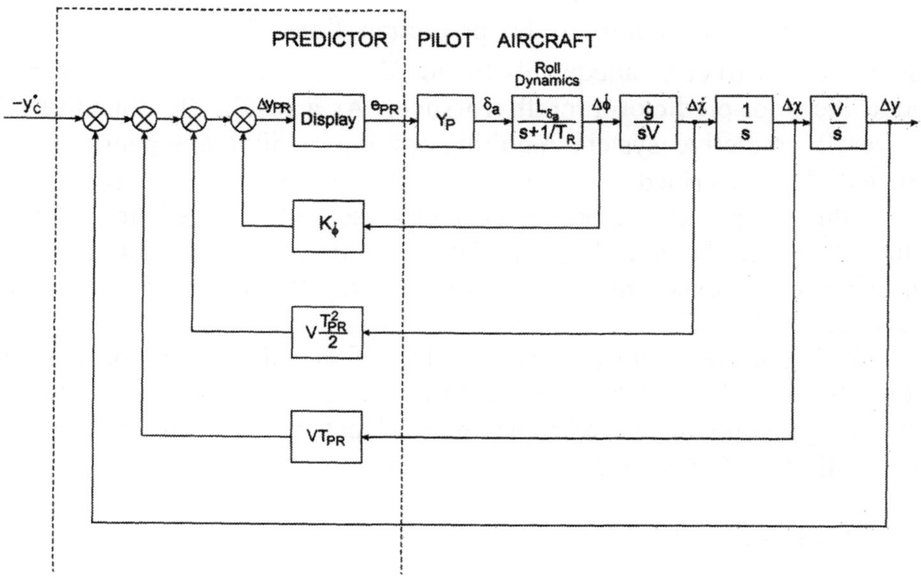

Fig. 15. Block diagram for lateral predictor with roll rate feedback.

$$Y_C = \frac{\Delta\phi(s)}{\delta_a(s)} = \frac{L_{\delta_a}}{s(s+1/T_R)}, \tag{4}$$

the following relation for the predictor-aircraft transfer function is obtained:

$$Y_{PR}Y_C = K_C^* \frac{(s^2 + 2\zeta_{PR}\omega_{PR}s + \omega_{PR}^2)}{s^3}, \tag{5}$$

where

$$K_C^* = (g/2)K_{PR}L_{\delta_a}T_{PR}^2T_R, \tag{6a}$$

$$\omega_{PR} \approx \sqrt{2}/T_{PR}, \tag{6b}$$

$$\zeta_{PR} \approx 1/\sqrt{2}. \tag{6c}$$

From Eq. (5), it follows that there is a K/s frequency region above ω_{PR}. By proper selection of the prediction time $T_{PR} \approx \sqrt{2}/\omega_{PR}$, it is possible to construct an adequately broad K/s frequency region centered around the pilot-predictor-aircraft crossover. As a result, the objective of an overall predictive system requiring minimum pilot compensation is achieved. The described K/s properties are illustrated in Fig. 16, which shows the frequency response characteristics of a predictor-aircraft system. The data shown in Fig. 16 relate to an aircraft used in pilot-in-the-loop simulation experiments; the relevant results are presented in a subsequent section.

A further issue is closed-loop stability of the pilot-predictor-aircraft system. In Fig. 17, the stability properties are evaluated with the root locus technique yielding results of rather general nature. The following pilot model valid for K/s characteristics is applied:

$$Y_P(s) = K_P\, e^{-\tau_e s}. \tag{7}$$

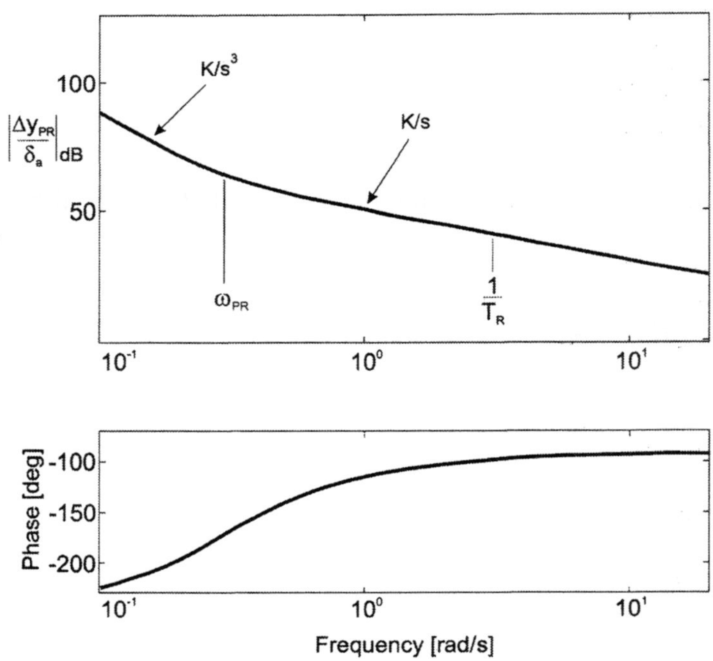

Fig. 16. Frequency response characteristics of predictor-aircraft system.

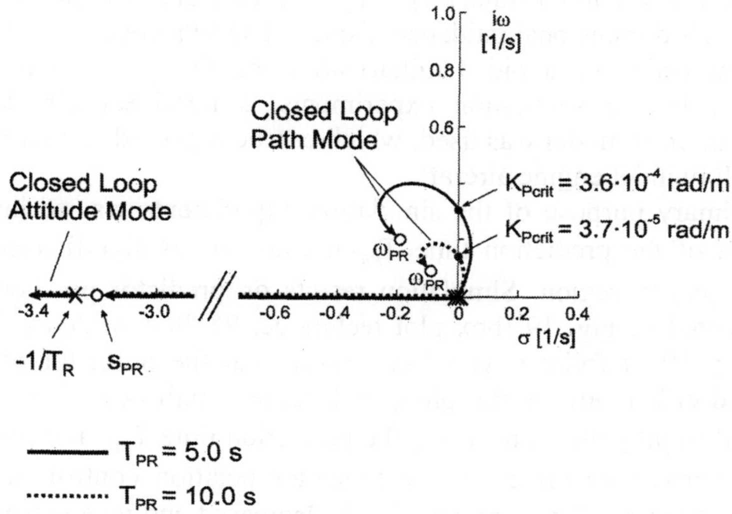

Fig. 17. Root locus of closed-loop system with roll rate feedback [$T_R = 3.0$ s , pilot model Eq. (7) with $\tau_e = 0.25$ s].

Basically, Fig. 17 shows that the system is stable for pilot gains above a certain value $K_{P_{crit}}$. Since the gain for pilot-system crossover is significantly greater than $K_{P_{crit}}$, it follows from the root locus result that the pilot-predictor-aircraft system is stable. Furthermore, Fig. 17 shows that there are basically two closed-loop modes, one primarily related to path and the other to attitude motions.

5. Results of Simulation Experiments for Three-Dimensional Guidance Display with Predictor

An experimental investigation of the described 3-dimensional guidance display with predictor was the subject of pilot-in-the-loop simulation tests. Five pilots with different professional background (airline pilots, private pilot, student pilot) performed the simulation experiments which were carried out at a fixed-base simulator. The layout of the 3-dimensional guidance display developed for the experimental program corresponds to the configuration shown in Fig. 13. The tasks of the pilot was to follow a curved trajectory (Fig. 18), indicated as command flight path in the 3-dimensional guidance display. The sequence of the turns was altered in order to avoid familiarization of the pilots with a fixed trajectory. In the simulation experiments, a nonlinear six degree-of-freedom aircraft model was used, which can be regarded as representative of a small twin jet engine aircraft.

A primary purpose of the simulation experiments was to investigate the effect of the prediction time T_{PR} because of its significance for the K/s frequency region. Simulation results on predictor position control are presented in Fig. 19 (box plot technique, 95 % confidence interval). From Fig. 19, it follows as a basic result that the predictor position is controlled effectively by the pilot, with rather small deviations from the command flight path. Concerning the prediction time T_{PR}, it turns out that it has a substantial effect on the predictor position control, showing a decrease of the predictor error as T_{PR} is decreased and vice versa. Control activity results are depicted in Fig. 20 which shows the correcting aileron

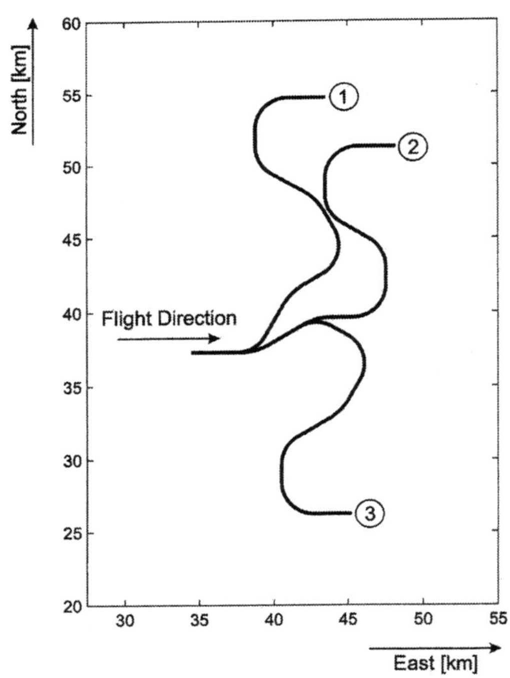

Fig. 18. Curved trajectory shapes of simulation experiments.

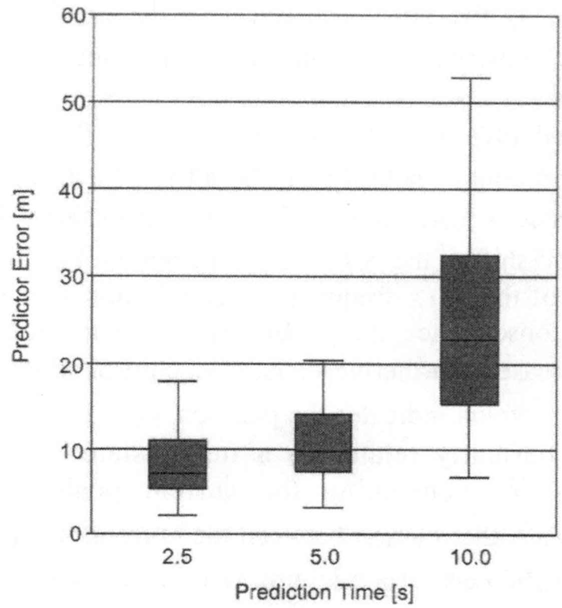

Fig. 19. Control of predictor position.

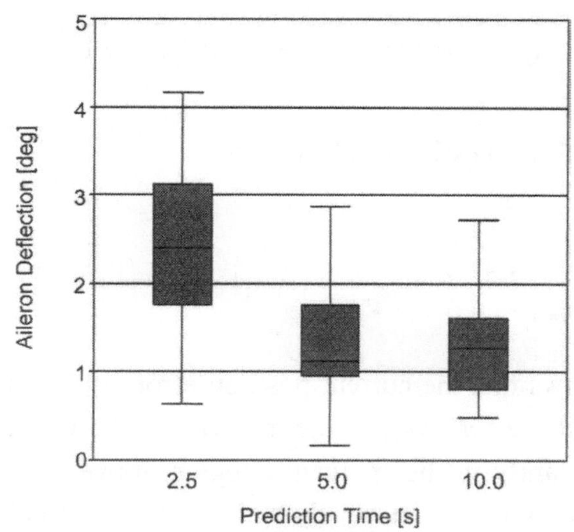

Fig. 20. Control activity of pilots.

commands given by the pilot. The effect of the prediction time T_{PR} is again significant, showing now an increase of the control activity as T_{PR} is decreased.

The described effects of the prediction time T_{PR} on the predictor position error and control activity can be attributed to pilot-loop closure behavior. Reference is made to Fig. 21, which shows that a decrease of T_{PR} yields a downward shift of the K/s frequency region. For loop closure, the downward shift of the K/s frequency region requires an increase of the pilot gain. As a consequence, the predictor position deviations are reduced when T_{PR} is decreased. Furthermore, the pilot control activity is increased.

The predictor, which indicates the position $y(t+T_{PR})$ at the prediction time ahead, is basically related to a future state. But it is also an efficient means for controlling the current position $y(t)$. Using $\Delta y(s) = g\Delta\phi(s)/s^2$, the relation between the current position error $\Delta y(s)$ and the future position error (predictor error) $\Delta y_{PR}(s) = (1/K_{PR})e_{PR}(s)$ can be expressed as

$$\frac{\Delta y(s)}{\Delta y_{PR}(s)} = \frac{1}{(T_R s+1)[(s/\omega_{PR})^2 + 2\zeta_{PR} s/\omega_{PR} + 1]}. \tag{8}$$

Accounting for $\zeta_{PR} = 1/\sqrt{2}$, it follows that

$$\left|\frac{\Delta y(s)}{\Delta y_{PR}(s)}\right|_{s=i\omega} \leq 1. \tag{9}$$

This relation shows that the current position error Δy is basically smaller than the predictor error Δy_{PR}. The reduction of Δy relative to Δy_{PR} increases significantly in the frequency region above ω_{PR}. Furthermore, both errors approach zero in steady-state reference conditions. This is because

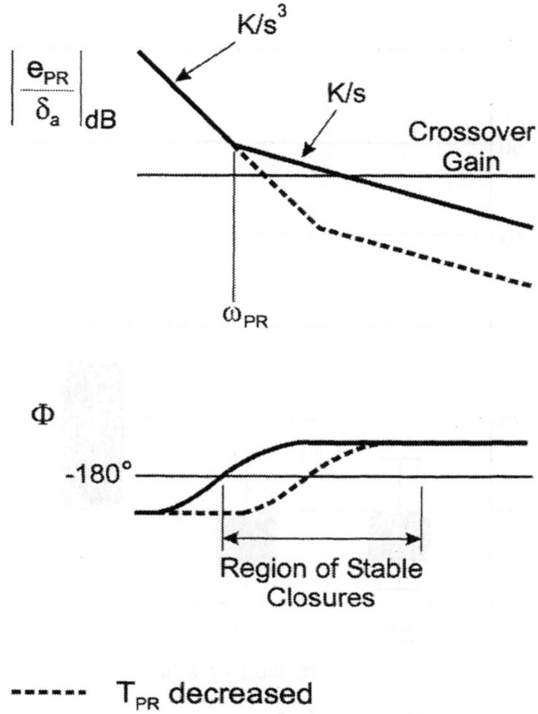

Fig. 21. Asymptotic Bode plot for predictor-aircraft system.

$$\left. \frac{\Delta y(s)}{\Delta y_{PR}(s)} \right|_{s \to 0} = 1. \tag{10}$$

These results were confirmed by pilot-in-the-loop simulation experiments. This is illustrated in Fig. 22, which shows that the deviations of the current position are smaller than those of the predictor position as depicted in Fig. 19.

6. Conclusions

A guidance display is considered providing the pilot with status and command information in a 3-dimensional format for current and future

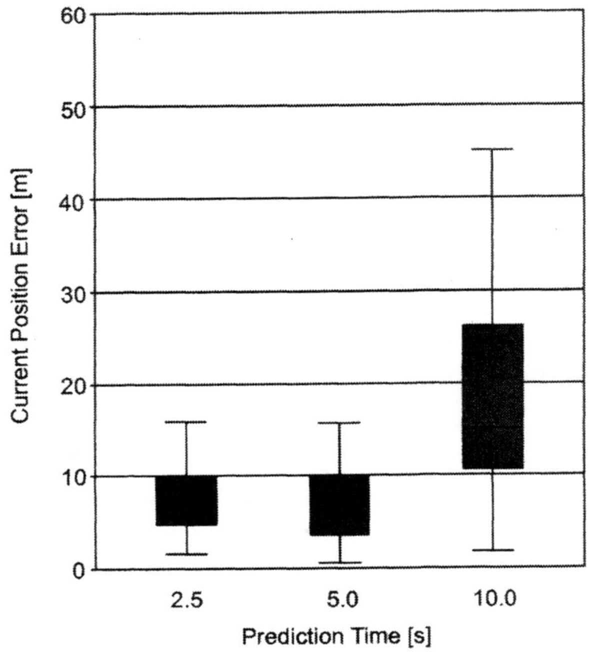

Fig. 22. Control of current position.

flight situations. The basic concept features a computer-generated imagery of the command flight path, other guidance information, and the outside world. The required attitude and position data for a correct adjustment of the displayed imagery are transferred from a precision navigation system using differential global positioning and inertial sensor data. A series of flight tests aiming at a wide range of applications of the 3D-guidance display were performed, with demanding control tasks for the pilots like precision approach and landing, low-level flights in highly curved, narrow river valleys, curved/steep/short approaches and low-level and terrain-following flights in mountainous areas. The flight test results show that the pilot controlled precisely the aircraft and held it on the command trajectory.

An extended display concept for presenting guidance information in a 3-dimensional format features a predictor which indicates the future position of the aircraft at a specified time ahead. For best results in terms of performance and workload, the predictive system should be designed such that the controlled predictor-aircraft element requires minimum pilot compensation. A predictor control law is developed for achieving this objective. Results from pilot-in-the-loop simulation experiments concerning significant predictor law parameters were performed, yielding verification of the design considerations.

References

1. THEUNISSEN, E., *Integrated Design of a Man-Machine Interface for 4D-Navigation*, PhD Thesis, Delft University of Technology, Delft, Netherlands, 1997.

2. THEUNISSEN, E., and MULDER, M., *Availability and Use of Information in Perspective FlightPath Displays*, Proceedings of the AIAA Flight Simulation Technologies Conference, pp. 137-147, 1995.

3. GRUNWALD, A.J., ROBERTSON, J.B., and HATFIELD, J.J., *Experimental Evaluation of a Perspective Tunnel Display for Three-Dimensional Helicopter Approaches*, Journal of Guidance, Control, and Dynamics, Vol. 4, No. 6, pp. 623-631, 1981.

4. GRUNWALD, A.J., *Tunnel Display for Four-Dimensional Fixed-Wing Aircraft Approaches*, Journal of Guidance, Control, and Dynamics, Vol. 7, No. 3, pp. 369-377, 1984.

5. GRUNWALD, A.J., *Predictor Laws for Pictorial Flight Displays*, Journal of Guidance, Control, and Dynamics, Vol. 8, No. 5, pp. 545-552, 1985.

6. GRUNWALD, A.J., *Improved Tunnel Display for Curved Trajectory Following: Control Considerations*, Journal of Guidance, Control, and Dynamics, Vol. 19, No. 2, pp. 370-377, 1996.

7. GRUNWALD, A.J., *Improved Tunnel Display for Curved Trajectory Following: Experimental Evaluation*, Journal of Guidance, Control, and Dynamics, Vol. 19, No. 2, pp. 378-384, 1996.

8. HASKELL, I.D., and WICKENS, C.D., *Two- and Three-Dimensional Displays for Aviation: A Theoretical and Empirical Comparison*, International Journal of Aviation Psychology, Vol. 3, No. 2, pp. 87-109, 1993.

9. WICKENS, C.D., FADDEN, S., MERWIN, D., and VERVERS, P.M., *Cognitive Factors in Aviation Display Design*, Proceedings of the 17th AIAA/IEEE/SAE Digital Avionics Systems Conference, Bellevue, Washington, 31 October – 6 November 1998, 0-7803-5086-3/98, 1998.

10. HELMETAG, A., MAYER, U., and KAUFHOLD, R., *Improvement of Perception and Cognition in Spatial Synthetic Environment*, Proceedings of the 17th European Annual Conference on Human Decision Making and Manual Control, Valenciennes, France, 14-16 December 1998, pp. 207-214, 1998.

11. LENHART, P.M., PURPUS, M., and VON VIEHBAHN, H., *Flight Testing of Cockpit Displays with Sinthetic Vision*, Yearbook 1998-I, German Society of Aeronautics and Astronautics, pp. 707-713, 1998 (in German).

12. FUNABIKI, K., MURAOKA, K., TERUI, Y., HARIGAE, M., and ONO, T., *In-Flight Evaluation of Tunnel-in-the Sky Display and Curved Approach Pattern*, Proceedings of the AIAA Guidance, Navigation, and Control Conference, pp. 108-114, 1999.

13. MULDER, M., *Cybernetics of Tunnel-in-the-Sky Displays*, Delft University Press, Delft, Netherlands, 1999.

14. SACHS, G.; and MÖLLER, H., *Synthetic Vision Flight Tests for Precision Approach and Landing*, Proceedings of the AIAA Guidance, Navigation, and Control Conference, pp. 1459-1466, 1995.

15. SACHS, G., DOBLER, K., and HERMLE, P., *Flight Testing Synthetic Vision for Precise Guidance Close to the Ground*, Proceedings of the AIAA Guidance, Navigation, and Control Conference, pp. 1210-1219, 1997.

16. SACHS, G., DOBLER, K., and THEUNISSEN, E., *Pilot-Vehicle System Control Issues for Predictive Flight Path Displays*,

Proceedings of the AIAA Guidance, Navigation, and Control Conference, pp. 574-582, 1999.

17. SACHS, G., *Flight Path Predictor for Minimum Pilot Compensation,* Aerospace Science and Technology, Vol. 3, No. 4, pp. 247-257, 1999.

18. SACHS, G., *Perspective Predictor/Flight Path Display with Minimum Pilot Compensation,* Journal of Guidance, Control, and Dynamics, Vol. 23, No. 3, pp. 420-429, 2000.

19. DITTRICH, J., KÜHMSTEDT, E., LECHNER, W., et al, *Experiments with Real Time Differential GPS Using a Low Frequency Transmitter in Mainflingen, Germany: Results and Experiences,* Paper Presented at EURNAV-94 Land Vehicle Navigation, Dresden, Germany, 14-16 June, 1994.

20. VIEWEG, S., and SCHÄNZER, G., *Precise Flight Navigation by Integration of Satellite Navigation Systems with Inertial Sensors,* Yearbook 1992-I, German Society of Aeronautics and Astronautics, pp. 171-177, 1992.

21. MCRUER, D.T., *Pilot Modeling,* AGARD Publication LS-157, Chapter 2, pp.1-30, 1988.

22. HESS, R. A., *Feedback Control Models: Manual Control and Tracking,* Handbook of Human Factors and Ergonomics, 2nd Edition, Edited by G. Salvendy, Wiley, New York, NY, pp. 1249-1294, 1997.

5

Neighboring Vehicle Design for a Two-Stage LaunchVehicle[1]

K. H. WELL[2]

Abstract. The paper presents numerical results of a study concerned with the simultaneous optimization of the ascent trajectory of a two-stage launch vehicle and some significant vehicle design parameters. Besides the trajectory design, models are given that relate (i) the propulsion mass to a desirable increase in the mass flow for the rocket engines and (ii) the structural mass of the fuel tanks to a desirable increase in the propellant mass. Using these models, it is shown how the example vehicle should be modified in order to carry a higher payload into an Earth escape orbit. It is shown that an overall increase of the vehicle liftoff mass of about 4% will result in a payload increase of about 11%.

Key Words. Trajectory optimization, launch vehicles, concurrent engineering.

---

[1] The author gratefully acknowledges the financial support provided by the European Space Technology Center (ESTEC) through its Contract Monitor Klaus Mehlem.

[2] Professor and Director, Institute of Flight Mechanics and Control, University of Stuttgart, 70550 Stuttgart, Germany.

1. Introduction

Reference 1 gives an overview of a trajectory optimization software (ASTOS) which has been developed over the past ten years for the European Space Agency. This software enables a user to specify a particular launch or reentry vehicle and a particular mission solely by data. It generates an initial estimate for the solution automatically, and it assists the user in the solution process via a user interface. Among many other features, it contains a particular capability which links the vehicle design to trajectory optimization and allows the combined optimization of the trajectory and the vehicle parameters. The purpose of this paper is to demonstrate this capability taking as an example the ascent of a two-stage launch vehicle into an Earth escape orbit, while simultaneously answering the question of how the nominal vehicle should be modified in order to increase the payload in that orbit.

Traditionally, vehicle design is mostly separated from atmospheric trajectory optimization. At most, atmospheric trajectories are simulated using particular guidance laws during the design process. However, in recent years, attempts have been made to link the task of finding the best ascent trajectory to the task of designing the vehicle size.

Reference 2 presents a design tool to this end. There, the optimization is organized hierarchically: The design optimization is performed in an outer loop; the trajectory optimization is performed in an inner loop. In the outer loop, the trajectory is frozen; in the inner loop, the design is frozen. The software has been applied successfully to reentry vehicle design as well as to design modifications of a winged launcher with air breathing propulsion.

In principle, the design process must take into consideration that, when changing the geometry of the vehicle, not only the mass data change but in particular the aerodynamic data do.

Therefore, once a particular change in geometry has occurred, appropriate aerodynamic methods have to be used to recalculate the aerodynamic coefficients. Depending on the required accuracy, more or less sophisticated aerodynamic codes have to be used which may lead easily to rather large amounts of computing times.

In this paper, it is assumed that the modifications from a reference design are small enough such that a recalculation of the aerodynamic coefficients is not needed. In addition, the diameter of the cylindrical vehicle stages is kept constant. This leads to the assumption that the drag forces will not be affected by the modifications. The essential design modifications are changes in the engine masses and tank sizes for the two stages. By limiting the modifications to 20% from the nominal values, it is conjectured that the results are realistic and can serve as guidelines for eventual modifications.

2. Reference Vehicle

Figure 1 shows the reference vehicle. It consists of two main stages,

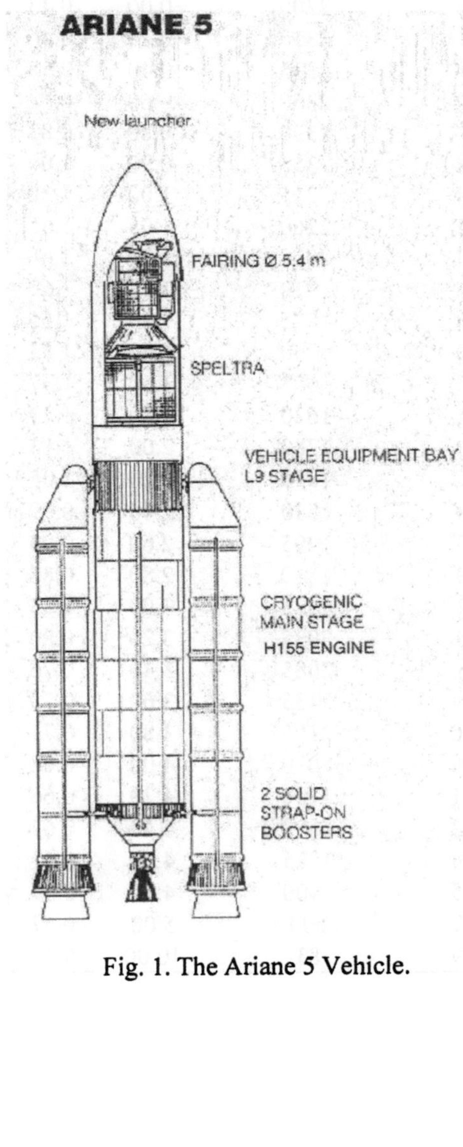

Fig. 1. The Ariane 5 Vehicle.

the lower cryogenic stage with the H155 engine and the two boosters P230 and the upper stage with the L9 engine. Figure 1 shows the version of the vehicle carrying two payloads; SPELTRA is the device which holds and separates the two payloads once orbital target conditions have been achieved by the upper stage. Table 1 contains the mass flow for a P230 booster and the drag coefficient of the vehicle (see also Fig. 2).

Table 1. Mass flow profile of a P230 booster and drag coefficient of the vehicle.

Time [s]	Mass flow [kg/s]	Mach	C_A
0	2200	0.00	0.43
6	2310	0.20	0.46
9	2335	0.40	0.53
12	2360	0.60	0.64
16	2390	0.80	0.83
18	2355	0.90	1.00
20	2315	1.00	1.30
21	2260	1.05	1.65
23	2000	1.10	1.68
25	1910	1.15	1.67
28	1775	1.20	1.65
30	1700	1.40	1.53
33	1655	1.60	1.37
36	1620	1.80	1.23
50	1780	2.00	1.13
55	1855	2.20	1.05
58	1830	2.40	0.99
59	1795	2.60	0.93
60	1810	2.80	0.88
62	1825	3.00	0.83
80	2035	3.20	0.80
85	2085	3.40	0.76
96	2185	3.60	0.73
100	2195	3.80	0.71
108	2010	4.00	0.68
111	1700	4.20	0.66
112	1690	4.40	0.64
114	1525	4.60	0.61
115	1400	4.80	0.59
122	500	5.00	0.57
129	83	10.00	0.57

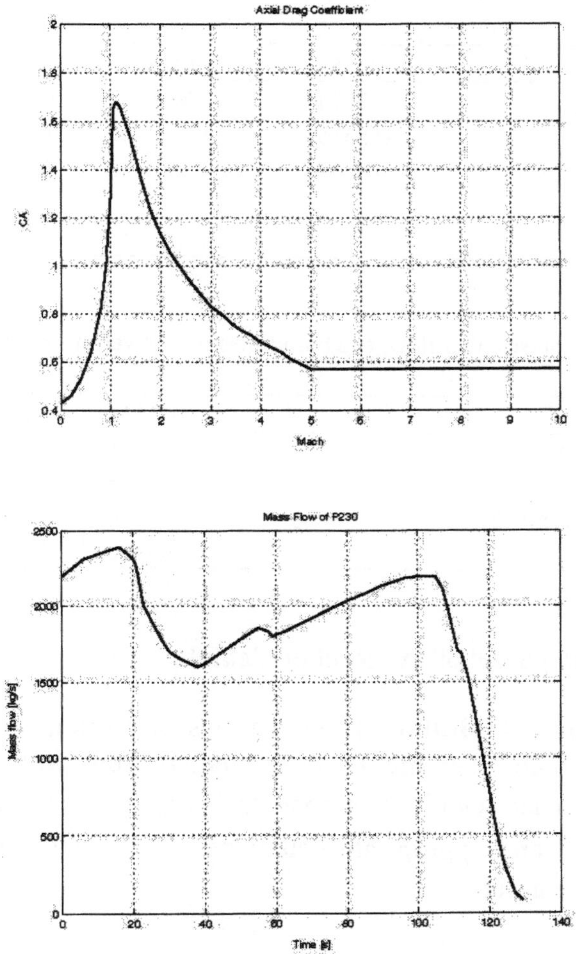

Fig. 2. Axial drag coefficient of the overall vehicle and mass flow for one booster.

Table 2 contains the masses of the various vehicle stages. Included in the total structural mass are 1401 kg for the vehicle equipment bay which is attached to the L9 stage and 1935 kg for the payload fairing which is ejected after burnout of the main stage. Table 3 gives the engine data, the x designating that these data are for experimental engines. About 0.8% of the fuel (unburned propellant) for both the H155x engines and the L9x engines cannot be utilized in the combustion process and thus does not contribute to the propulsion of these two engines.

Table 2. Masses for the reference vehicle.

	H155x	2P230	L9x	Sum
Structure [kg]	13778	78188	1222	96524
Propellant [kg]	155315	472226	9381	636922
Engine [kg]	1672.35		28.23	1701
Payload [kg]				3147
Launch mass [kg]				738293

Table 3. Specific impulses, fuel flow, and burn times for the various propulsion systems.

	H155x	P230	L9x
Vacuum specific impulse [s]	428.68	270.20	323.20
Nominal mass flow [kg/s]	267.78	Table 1	9.20
Nominal nozzle area [m**2]	2.50	6.35	1.13
Burn time [s]	580.00	129.10	open

3. Mathematical Model of the Rocket Vehicle

3.1. Equations of Motion. The equations of motion of the center of mass over an oblate, rotating Earth are taken from Ref. 3. The state variables are: Inertial velocity components V_λ, V_δ, V_R (see Fig. 3), position variables R, λ, δ, and appropriate equations for the mass change of the vehicle during the ascent

$$\frac{d}{dt}\begin{bmatrix} R \\ \lambda \\ \delta \end{bmatrix} = \begin{bmatrix} V_R \\ \dfrac{V_\lambda}{R\cos\delta} - \Omega_E \\ \dfrac{V_\delta}{R} \end{bmatrix}, \tag{1a}$$

$$\frac{d}{dt}\begin{bmatrix} V_R \\ V_\lambda \\ V_\delta \end{bmatrix}_L = \frac{1}{R}\begin{bmatrix} V_\lambda^2 + V_\delta^2 \\ V_\lambda\left(V_\delta \tan\delta - V_R\right) \\ -V_\lambda^2 \tan\delta - V_R V_\delta \end{bmatrix}_L + \frac{\vec{F}}{m}, \tag{1b}$$

where Ω_E is the Earth rotational speed and $\vec{F}$ is the vectorial sum of all the external forces. The subscript L indicates inertial variables in the local horizontal coordinate system. The external forces are the thrust forces

$$\vec{T} = \begin{bmatrix} g_0 I_{spvac} \dot{m}_b(t) - A_e p \\ 0 \\ 0 \end{bmatrix}_B, \tag{2}$$

the aerodynamic forces

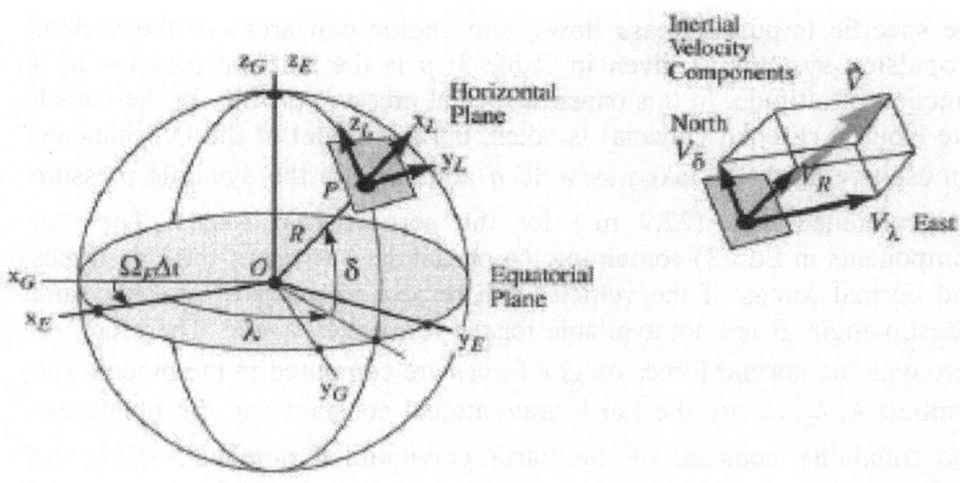

Fig. 3. Definition of the planet-relative position and inertial velocity state variables. The inertial geocentric axis system (subscript G) and the rotating planet-fixed axis system (subscript E) span the equatorial plane, their z axes coinciding with the planet axis of rotation. The local horizontal axis system (subscript L) is attached to the vehicle center of mass.

$$\bar{F}_{aero} = qA_{ref} \begin{bmatrix} -C_A(M) \\ \left(\frac{\partial C_Y}{\partial \beta}(M) \right) \beta \\ -\left(\frac{\partial C_N}{\partial \alpha}(M) \right) \alpha \end{bmatrix}_B \tag{3}$$

and the gravitational forces

$$\bar{G} = \frac{-k}{R^2} \begin{bmatrix} 1 + \frac{3J_2}{2}\left(\frac{R_E}{R}\right)^2 (1 - 3\sin^2\delta) + 2J_3\left(\frac{R_E}{R}\right)^3 (3 - 5\sin^2\delta)\sin\delta \\ 3J_2\left(\frac{R_E}{R}\right)^2 \sin\delta\cos\delta + \frac{3J_3}{2}\left(\frac{R_E}{R}\right)^3 (5\sin^2\delta - 1)\cos\delta \end{bmatrix}_L . \tag{4}$$

The thrust force for the boosters is a function of time as given in Table 1; g_0 is the gravitational acceleration at sea level; $I_{spvac}, \dot{m}_b(t)$, and A_e are the specific impulses, mass flows, and engine exit areas of the various propulsion systems as given in Table 3; p is the ambient pressure as a function of altitude. In this paper, a special pressure profile for the launch site Kourou (French Guyana) is taken, but the model of the US Standard atmosphere might be taken as well; q and A_{ref} are the dynamic pressure and reference area (22.9 m^2) for the aerodynamic forces. The two components in Eq. (3) containing the partial derivatives of the side forces and normal forces of the vehicle with respect to angle of attack α and sideslip angle β are not available for the reference vehicle. Therefore, no aerodynamic normal forces or side forces are computed in the model. The symbols k, J_2, J_3 are the Earth gravitational constant and the oblateness and triaxiality constant of the Earth gravitational potential, R_E is the equatorial radius of the Earth; their values are given, for example, in Ref. 3. The subscripts B in equations (2) and (3) describe the forces in body axes. To transform them into the local horizontal axes, the transformation relations

$$\bar{T}\big|_L = T_1(\Psi)T_2(\pi/2 - \Theta)T_1(\Phi)\bar{T}\big|_B \tag{5}$$

are to be applied where Ψ, Θ, Φ are the azimuth, pitch, and roll angles describing the launchers attitude with respect to the local horizontal system. The transformation matrices are

$$T_1(\varepsilon) = \begin{bmatrix} 1 & 0 & 0 \\ 0 & \cos\varepsilon & \sin\varepsilon \\ 0 & -\sin\varepsilon & \cos\varepsilon \end{bmatrix}, \; T_2(\varepsilon) = \begin{bmatrix} \cos\varepsilon & 0 & -\sin\varepsilon \\ 0 & 1 & 0 \\ \sin\varepsilon & 0 & \cos\varepsilon \end{bmatrix}. \quad (6)$$

It is assumed that the vehicle will not roll during launch and, therefore, $\Phi = 0$ is set. The same transformation applies to the aerodynamic forces in (3). With these definitions, the controls for the ascent problem are the pitch and yaw angles.

3.2. Mass Models for Engine and Tank Sizing. For the task at hand, it is assumed that the boosters are given and are not to be modified. Modifiable are the two rocket engines and the size of the tanks for the fuel of the main and the upper stages. By scaling the engine up, the mass flow, nozzle area, and thrust of a particular engine can be increased causing, of course, an increase in the engine mass. Simple models describing the interrelation of these data are taken from Ref. 4. Given a sizing parameter $z_{\min} \le z \le z_{\max}$, the mass flow is modeled as

$$\dot{m}_b(\sigma) = \dot{m}_0(\sigma) z, \quad (7)$$

where $\dot{m}_0(\sigma)$ is the reference mass flow and $0 \le \sigma \le 1$ is the throttle setting (an additional control). The nozzle exit area is

$$A_e = A_0 z; \quad (8)$$

the maximum thrust is

$$T_{\max} = g_0 I_{spvac} m_b(1) - A_e p(h), \quad (9a)$$

the actual thrust itself is

$$T = T_{\max} \sigma. \quad (9b)$$

With these definitions, the engine mass can be calculated as

$$m_e = a[T_{\max}(z)]^b, \tag{10}$$

where a, b are correlation coefficients. Similarly for tank sizing, the propellant mass and the structural mass are

$$\mathrm{m}_p = \mathrm{m}_{p0} z, \tag{11}$$

$$m_s = a\left[m_p\right]^b, \quad a = m_{s0} / (m_{p0})^b. \tag{12}$$

Here, m_{p0} is the reference value for the propellant mass and m_{s0} is the reference value for the structural mass; b is another correlation coefficient. As mentioned in the introduction, the tank size is to be varied assuming a constant diameter. Then, the change in tank volume is computed from

$$\Delta V = \frac{m_{p0}(z+1)}{\rho_p} \tag{13}$$

with ρ_p as mean density of the fuel. Assuming a cylindrical shape of the tank, this results in a change in length of the tank,

$$\Delta l = \frac{4}{\pi d^2} \Delta V. \tag{14}$$

Tables 4 and 5 present the data used in the subsequent calculations. Figure 4 shows how the dry engine masses change with increasing fuel flow and how the tank masses change with increasing amounts of fuel. Altogether there are four design parameters $z_i, i = 1 \ldots 4$, to be chosen in the optimization.

Table 4. Engine sizing parameters.

	H155x	L9x
Engine sizing factor	1.00	1.00
Mass flow [kg/s]	267.78	9.20
Nozzle area [m**2]	2.50	1.13
Engine mass [kg]	1672.35	28.23
Vacuum thrust [kN]	1125.74	29.15
a	3.62	0.89
b	0.87	1.03

Table 5. Tank sizing parameters.

	Main stage	Upper stage
Length [m]	29.00	4.50
Diameter [m]	5.40	5.40
Tank sizing factor	1.00	1.00
a	0.24	1.65
b	0.92	0.72
Mean density [kg/m**3]	280.00	1260.00

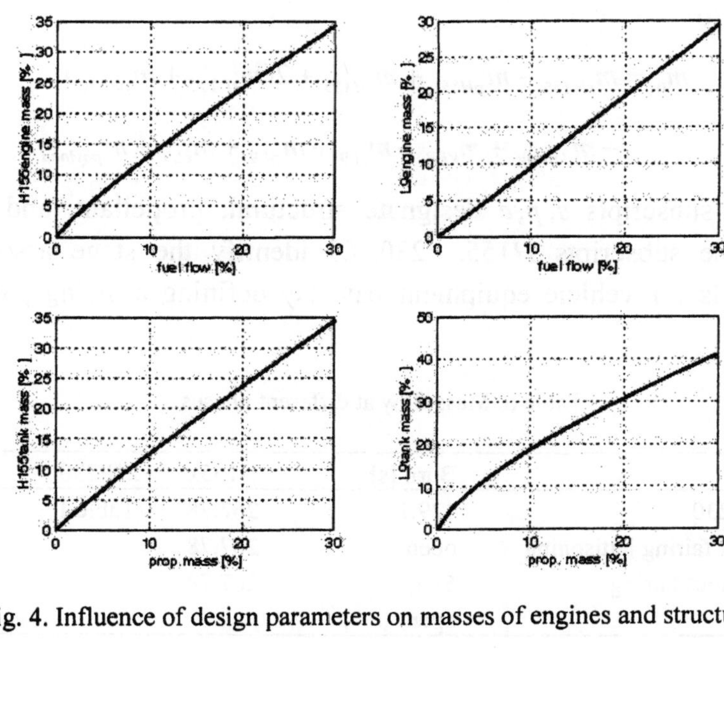

Fig. 4. Influence of design parameters on masses of engines and structure.

4. Multiphase Optimal Control Problem

The main constraints of the trajectory optimization problem are the differential system described by equations (1)-(6) and the additional differential equation for the change in mass,

$$\dot{m} = -m_{b_i}, \tag{15}$$

where m_{b_i} is the fuel flow in the ith phase, $i = 1...4$. Table 6 contains the mass flow data for each phase. The first phase consists of the simultaneous burn of the main engine and the two boosters, the second and third phase are with the main engine only, and the fourth phase is the burning of the upper stage engine. The booster burn time is fixed, so is the overall burn time of the main engine. The times for the fairing jettisoning and for the L9 engine cut-off are kept free in the optimization process.

4.1. Initial Conditions. The vehicle is supposed to be launched from Kourou (French Guyana). The initial values for the position are altitude $(+R_E)$, geographical longitude and latitude of Kourou. The initial velocity components are taken to be $(5,0,0)^T$. The liftoff mass is computed according to

$$m_0 = m_{sH155} + m_{pH155} + m_{eH155} + 2(m_{sP230} + m_{pP230})$$

$$+ m_{fairing} + m_{VEB} + m_{sL9} + m_{pL9} + m_{eL9} + m_{payload}. \tag{16}$$

Here, the subscripts s, p, e designate structural, propellant, and engine masses, the subscripts $H155$, $P230$, $L9$ identify the stage association. VEB stands for vehicle equipment bay. By defining a sizing parameter

Table 6. Mass flow at different phases.

	Burn [s]	H155x	2*P230	L9x
H155x+2P230	129.1	267.78	Table 1	
H155x until fairing jettisoning	open	267.78		
H155x without fairing	580	267.78		
L9x	open			9.20

according to Section 3.2 for the amount of propellant to be used in each of the main stage and the upper stage as well as in each of the engines, there are altogether five parameters to be optimized, the fifth being the payload mass $m_{payload}$. In this way, the initial mass is a function of these five parameters.

4.2. Target and Intermediate Conditions, Cost Function. In order to define the final boundary conditions, a few auxiliary variables need to be defined (see Ref. 3). The inertial path inclination, inertial azimuth, and velocity are computed as

$$\tan\gamma_I = \frac{V_R}{\sqrt{V_\lambda^2 + V_\delta^2}}, \quad \tan\chi_I = \frac{V_\lambda}{V_\delta}, \quad V = \sqrt{V_R^2 + V_\lambda^2 + V_\delta^2}. \tag{17}$$

Furthermore, with $E = E_{kin} + E_{pot} = \frac{1}{2}V^2 - \frac{k}{R}$, the parameter f is defined as

$$f = \frac{2E_{kin}}{|E_{pot}|} = \frac{RV^2}{k}, \tag{18}$$

and the semimajor axis and eccentricity are defined as

$$a = \frac{R}{2-f}, \qquad e = \sqrt{1 - f(2-f)(\cos\gamma_I)^2}. \tag{19}$$

These parameters take on different values for different kind of conic sections,

$$f < 2, \quad a > 0, \quad e < 1, \quad \text{elliptic orbit,} \tag{20}$$

$$f = 2, \quad a = \infty, \quad e = 1, \quad \text{parabolic orbit,} \tag{21}$$

$$f > 2, \quad a < 0, \quad e > 1, \quad \text{hyperbolic orbit.} \tag{22}$$

From orbital mechanics, it is known that the true anomaly can be computed from

$$e\sin\nu = f\sin\gamma_I\cos\gamma_I,\ \ e\cos\nu = f(\cos\gamma_I)^2 - 1. \tag{23}$$

The orbital elements i, Ω, ω can be calculated by applying the laws of spherical trigonometry to the triangle with the sides $\omega+\nu, \Lambda-\Omega$, δ in Fig. 5. Here, Λ is the inertial longitude of the vehicle at a particular time. It can be defined as

$$\Lambda = \lambda_{t=0} + \Omega_E t, \tag{24}$$

where $\lambda_{t=0}$ is the geocentric longitude at the time of launch. From this figure, one obtains

$$\cos i = \cos\delta\sin\chi_I, \tag{25}$$

$$\sin(\omega+\nu) = \frac{\sin\delta}{\sin i},\ \ \cos(\omega+\nu) = \cos\delta\cos(\Lambda-\Omega), \tag{26}$$

$$\sin(\Lambda-\Omega) = \frac{\tan\delta}{\tan i},\ \ \cos(\Lambda-\Omega) = \frac{\cos\chi_I}{\sin i}. \tag{27}$$

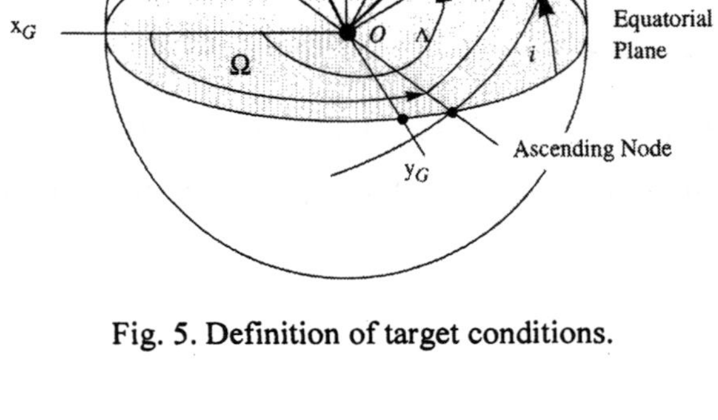

Fig. 5. Definition of target conditions.

Since the true anomaly υ is known from (23), these three equations can be used to determine the unknown orbital elements.

A hyperbolic target orbit can be defined by its excess velocity

$$V_e = \lim_{R\to\infty} V = \lim_{R\to\infty} \sqrt{\frac{2}{R} - \frac{k}{a}} = \sqrt{-\frac{k}{a}}, \tag{28}$$

its true anomaly for $R \to \infty$

$$\lim_{R\to\infty} \cos\nu = \cos(\nu_\infty) = \lim_{R\to\infty} \left(\frac{a(1-e^2)}{Re} - \frac{1}{e} \right) = -\frac{1}{e}, \tag{29}$$

and the declination of its asymptote for $R \to \infty$

$$\sin\delta_\infty = \sin i \sin(\omega + \nu_\infty) = \sin i \sin\left(\omega + \arccos\left(-\frac{1}{e}\right)\right). \tag{30}$$

By specifying the excess velocity and the declination of the asymptote, the semimajor axis and the inclination are defined via (28) and (25) for a given velocity vector, the parameter f and the eccentricity are calculated from (19) for a given value of R, and ν, ω, Ω are computed from (29), (26), and (27). As intermediate conditions, the perigee altitude

$$h_p = (1-e)a - R_E \tag{31}$$

and the heat flux

$$\dot{Q} = \frac{\rho}{2} V^3 \tag{32}$$

are needed.

Finally, the cost function for the optimal control problem is to maximize $m_{payload}$. Table 7 summarizes the trajectory optimization problem.

The design parameters are $\dot{m}_{bH155}, m_{pH155}, \dot{m}_{L9}, m_{pL9}, m_{payload}$ with the subscripted notation as described above. Of course, the integrals over the mass flow for each vehicle stage must satisfy the conditions

$$\int_{t_{ignition}}^{t_{burnout}} \dot{m}_{bH155} dt = m_{pH155}, \quad \int_{t_{ignition}}^{t_{burnout}} \dot{m}_{bL9} dt = m_{pL9}. \tag{33}$$

5. Solving the Trajectory Optimization Problem

The ASTOS software has been used to solve the above described problem. Inside the software, two methods are implemented; one is a direct multiple shooting method, first suggested in Ref. 5; the other method is based on direct collocation; see Ref. 6. Both methods transcribe the continuous optimal control problem into high parametric nonlinear programming problems which are solved by standard software. Inside ASTOS, two nonlinear programming solvers are implemented: A sequential linear least squares quadratic program solver (SLLSQP, Ref. 7) and a sparse nonlinear optimization solver (SNOPT, Ref. 8).

5.1. Initial Guess. In order to generate the initial time histories for the controls and the states, a guidance law based on the required velocity

Table 7. Trajectory optimization problem.

Phase	Duration [s]	Active propulsion	Attitude controls	Boundary constraints	Path constraints	Remark
1	0 - 129.1	H155X +2 P230	Ψ : linear law Θ : optimized		$\chi \leq 91.5°$	P230s are jettisoned at end
2	129.1 - t_{f2}	H155X	Ψ : optimized Θ : optimized	$\dot{Q} = 1135 \text{W/m}^2$		Payload fairing jettisoned at end
3	t_{f2} - 580.0	H155X	Ψ : optimized Θ : optimized	$h_p \leq 50$ km	$\dot{Q} \leq 1135 \text{W/m}^2$	H155X jettisoned at end
4	580.0 - t_{f4}	L9X	Ψ : optimized Θ : optimized	$V_\infty = 2.5$ km/s $\delta_{asymptote} = -12°$	$\dot{Q} \leq 1135 \text{W/m}^2$	

concept (Ref. 9) is used. Although in principle only applicable for elliptical target orbits, it can be used for the above problem as an approximation by choosing a sufficiently large apogee altitude for the makeshift target orbit.

For given orbital parameters r_a, r_p, i of such a highly eccentric target orbit, one computes the reference velocity, that is, that particular velocity which the vehicle should have in the desired orbit at that radius vector. The components in a local horizontal system are

$$\vec{V}_r = \begin{bmatrix} V_r \sin\gamma_r \\ V_r \cos\gamma_r \sin\chi_I \\ V_r \cos\gamma_r \cos\chi_I \end{bmatrix}, \tag{34}$$

where $V_r = \mu_e(2/r - 1/a)$ and

$$\chi_I = \arcsin(\cos i / \cos\delta), \tag{35}$$

$$V_r \cos\gamma_r = h/r, \tag{36a}$$

$$V_r \sin\gamma_r = \sqrt{k(2/r - 1/a) - (h/r)^2}, \tag{36b}$$

with

$$a = (r_a + r_p)/2, \qquad e = 1 - r_p/a, \tag{37a}$$

$$p = a(1 - e^2), \qquad h = \sqrt{kp}. \tag{37b}$$

γ_r, χ_I are the inertial elevation and the azimuth angles of the required velocity vector with respect to the local horizontal system; see Fig. 3. The difference between the required velocity vector $\vec{V}_r$ and the actual velocity vector $\vec{V}$ is

$$\vec{V}_d = \vec{V}_r - \vec{V}. \tag{38}$$

This velocity difference must ultimately be zero. One can show (see e.g. Ref. 10) that, by accelerating the vehicle in the direction of $\vec{V}_d$, this goal can be achieved.

According to Fig. 6, the vehicle acceleration is $\vec{a}_{veh}$. The direction and magnitude of Θ are obtained from

$$\vec{a}_t = \vec{g}_e + \vec{a}_{veh}, \tag{39}$$

where $g_e = \hat{V}^2/r - \mu_e/r^2$ is the effective gravitational acceleration, with $\hat{V}^2 = \hat{V}_\lambda^2 + V_\delta^2$. From the figure, one gets

$$\frac{\sin\beta}{a_{veh}} = \frac{\sin\varepsilon}{a_1}, \tag{40a}$$

$$a_t^2 = g_e^2 + a_{veh}{}^2 - 2g_e a\cos\varepsilon, \tag{40b}$$

and after some manipulations, as the solution of the quadratic equation

$$\Theta = \frac{\pi}{2} - \arccos\left(-\frac{P}{2} \pm \sqrt{\frac{P^2}{4} - Q}\right), \tag{41}$$

with

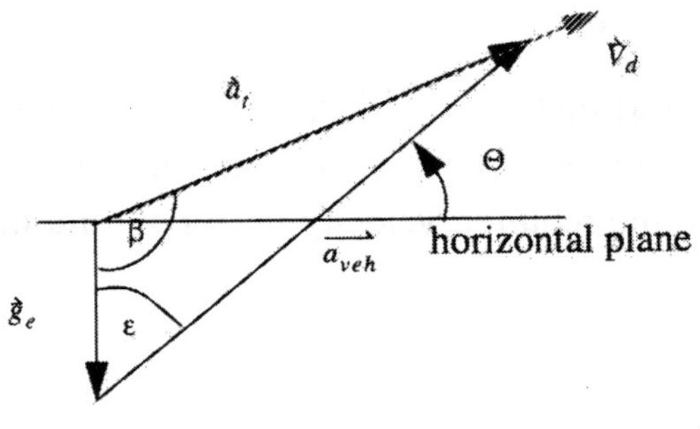

Fig. 6. Definition of the difference velocity vector $\vec{V}_d$.

$$P = -\frac{2g_e(\sin\beta)^2}{a_{veh}}, \tag{42a}$$

$$Q = (\sin\beta)^2\left(\left(\frac{g_e}{a_{veh}}\right)^2 + 1\right) - 1, \tag{42b}$$

where the appropriate sign has to be chosen. The yaw direction of the thrust vector is simply

$$\Psi = \chi_I. \tag{43}$$

With this guidance law, a nominal trajectory can be obtained by integrating the equations of motion from the initial state. The complete initial guess is obtained in the following three major steps,

Step 1: Vertical ascent followed by a constant pitch rate until a prescribed pitch attitude is obtained. Flight with this attitude until the angle of attack is zero.

Step 2: Gravity turn, that is, flight with zero angle of attack until some user specified event or time, usually until the burnout of the major stage.

Step 3: Guidance steering according to the above procedure until a specified time or until the desired orbit has been reached.

5.2. Optimal Solutions. Figure 7 shows the state time histories $R(t)$ and $V_R(t)$ as well as the control time histories $\Psi(t)$ and $\Theta(t)$ for both the initial guess and the nominal solution, that is, the solution with fixed values of the design parameters. These nominal values are given in Table 8 together with the optimal values. Figure 8 shows the altitude, ground track, inertial speed, and osculating perigee altitude of the H155x stage. These as well as other state and control time histories of the nominal case do not differ much from those of the optimal case. Both altitude and speed are somewhat smaller for a given time, due to the fact that the vehicle is heavier initially. Both trajectories have to satisfy the intermediate

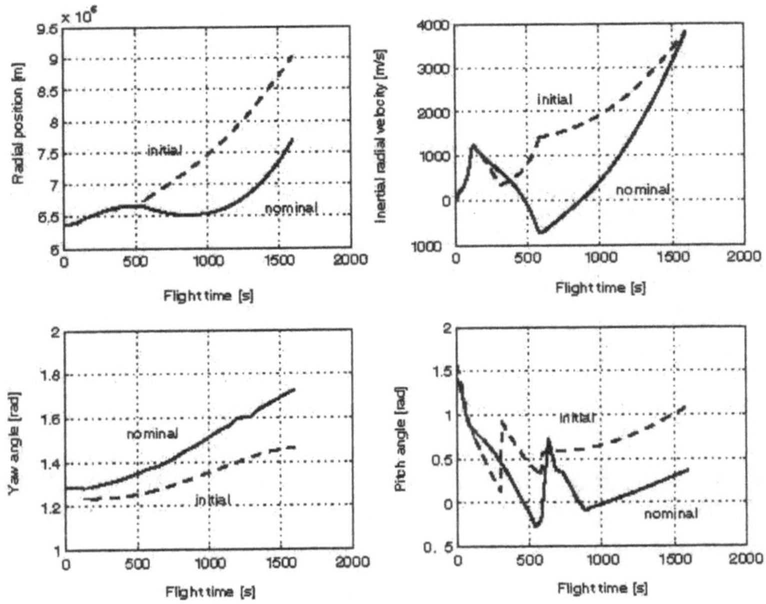

Fig. 7. Initial guess and nominal solution.

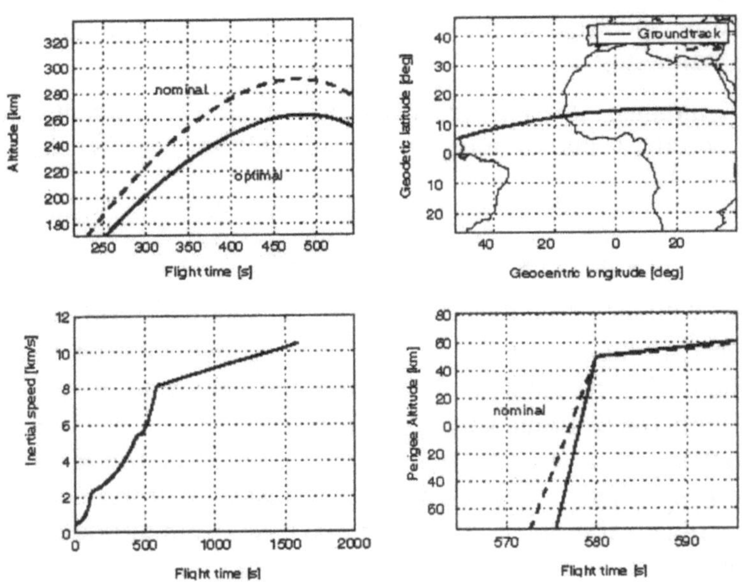

Fig. 8. Comparison between nominal and optimal solutions.

Table 8. Nominal and optimal values of the design parameters and associated model parameters.

	H155x	H155opt	2P230	L9x	L9opt	sum	sumopt
Structure [kg]	13778	15787	78188	1222	1394	96524	98705
Propellant [kg]	155315	180165	472226	9381	11257	636922	663649
Stage length [m]	29.00	33.84		4.50	4.57		
Engine [kg]	1672.35	1960.93		28.23	34.03	1701	1995
Mass flow [kg/s]	267.78	321.34		9.20	11.04		
Nozzle area [m**2]	2.5	3.0		1.13	1.36		
Vacuum thrust [kN]	1125.7	1350.9		29.15	34.98		
Payload [kg]						3147	3485
Launch mass [kg]						738293	767834
Structure%		14.6%			14.1%		2.3%
Propellant%		16.0%			20.0%		4.2%
Engine%		17.3%			20.6%		17.3%
Launch mass%							4.0%
Payload%							10.8%

boundary condition on the osculating perigee altitude at 580 sec. The overall flight time is approximately the same. The main difference between the nominal and the optimal case is shown in Table 8. The upper part of the table contains the structural, propulsive, and engine masses for both the nominal case and the optimal case. The VEB mass and fairing mass are included in the overall structural mass. The five design parameter values are given and can be compared to the nominal values. Due to the increased engine and fuel mass, the main stage needs to be extended by approximately 5m, while the geometric modifications of the upper stage are small. The lower part of the table gives the resulting changes in percent compared to the nominal design. The order of magnitude of the changes is between 15 to 20% for each stage; the changes are rather small for the vehicle altogether, since the booster mass contributes significantly to the overall mass of the vehicle. The overall increase in engine mass is 17% with respect to the engine masses without boosters. The increase in liftoff mass is about 4%; the increase in payload is about 11%.

6. Conclusions

The paper addresses the simultaneous optimization of both the ascent trajectory and some typical vehicle design parameters of a two-stage launch vehicle. It is shown that the modified vehicle does not influence the ascent trajectory to a great extent, which is not surprising, since for rocket propelled conventional launch vehicles the performance of the propulsion system depends weakly on the atmospheric conditions through the back pressure. This is due partly to the modeling assumptions that the diameter of the vehicle geometry has been held constant and that the dynamic lift of the vehicle has not been taken into account. By removing these restrictions, a greater interdependence between design and trajectory is conjectured to be observed. The approach presented here is applicable to launch vehicles with airbreathing propulsion as well where the interaction between design and trajectory is much more predominant.

References

1. WELL, K. H., MARKL, A., and MEHLEM. K., *ASTOS: A Trajectory Analysis and Optimization Software for Launch and Reentry Vehicles*, Paper IAF-97-V4.04, 48th International Astronautical Congress, Turin, Italy, 1997.

2. RAHN, M., SCHOETTLE, U. M., and MESSERSCHMID, E., *Multidisciplinary Design Tool for System and Mission Optimization of Launch Vehicles*, 6th AIAA/NASA/ISSMO Symposium on Multidisciplinary Analysis and Optimization, Bellevue, Washington, USA, 1996.

3. BUHL, W., EBERT, K., and WOLFF, H., *Technical Report 2, Modelling: Advanced Launcher Trajectory Optimization Software Technical Documentation*, European Space Technology and Research Center, Nordwijk, Netherlands, Contract 8046-88-NL-MAC, 1992.

4. SCHÖTTLE, U., and RAHN, U., *Fahrzeugmodelle für Sensitivitätsstudien konventioneller Trägerraketen (Vehicle Modelling of Conventional Launch Vehicles for Sensitivity Analysis)*, Institute for Space Systems, University of Stuttgart, Stuttgart, Germany, Report IRS 95-IB-11, 1995 (in German).

5. BOCK, H. G., and PLITT, K. J., *A Multiple Shooting Algorithm for Direct Solution of Optimal Control Problems*, Proceedings of the 9th IFAC World Congress, Budapest, Hungary, pp. 243-247, 1984.

6. HARGRAVES, C. R., and PARIS, S. W., *Direct Trajectory Optimization Using Nonlinear Programming and Collocation*, Journal of Guidance, Control, and Dynamics, Vol. 10, pp. 338-342, 1987.

7. KRAFT, D., *TOMP - FORTRAN Modules for Optimal Control Calculations*, VDI Fortschrittsberichte, Volume 8, No. 254, 1991.

8. GILL, P. E., MURRAY, W., and SAUNDERS, M. E., *Users Guide for SNOPT 5.3: A Fortran Package for Large-Scale Nonlinear Programming*, Department of Mathematics, University of California, San Diego, Report NA 97-5-4, 1997.

9. BATTIN, R., *Astronautical Guidance*, McGraw-Hill, New York, NY, 1964.

10. GRIMM, W., and WELL, K.H., *Guidance*, Lecture Notes, Institute for Flight Mechanics and Control, University of Stuttgart, 1994 (in German).

6

Controller Design for a Flexible Aircraft

M. HANEL[1] and K. H. WELL[2]

Abstract. The paper presents an overview of modeling the dynamic behavior of a large four-engine flexible aircraft and considers some of the options for control system design. The first part describes how to build an integral model, which can be used for simulating the rigid motion as well as the flexible motion of the aircraft. The result is a system of nonlinear equations of motion. The second part analyzes the dynamic properties of a sample aircraft by considering the linearized equations of motion for flight in a vertical plane at several operating points in the flight envelope. Here, it is shown how the eigenfrequencies of the rigid body and the elastic motion change with the load and flight conditions. In the third part, three options for control system design are discussed: (i) a conventional SAS controller, which does not influence actively the elastic behavior; (ii) an output feedback controller; and (iii) a robust controller. It is concluded that, using an integral controller, certain flying quality criteria can be met and damping of all the elastic modes can be improved.

Key Words. Flight control, aeroservoelasticity, flexible aircraft.

Research Scientist, Institute of Flight Mechanics and Control, University of Stuttgart, 70550 Stuttgart, Germany.

Professor and Director, Institute of Flight Mechanics and Control, University of Stuttgart, 70550 Stuttgart, Germany.

1. Introduction

The evolution of large transport aircraft is characterized by fuselages getting longer and wing spans getting wider, while efforts to reduce the structural weight reduce the structural stiffness. Both effects lead to more flexible aircraft structures with significant aeroelastic coupling between flight mechanics and structural dynamics, especially at high speed, high altitude cruise. This means that flight maneuvers and gusts may incite strong elastic reactions, which influence also the rigid-body flight mechanics.

Ride comfort and structural loads, especially for flight in a turbulent atmosphere, are influenced strongly by the vibrations of the aircraft structure. Since these vibrations cannot be controlled by conventional stability augmentation systems (SAS), some modern aircraft are equipped with additional control loops to improve the ride comfort (Ref. 1).

Stability augmentation and aeroelastic control loops are separated by dynamic filters. As rigid body dynamics and low frequency elastic modes get closer with increasing structural flexibility, the separate design of stability augmentation system and aeroelastic control loops becomes more difficult. Therefore, several recent studies (Refs. 2-5) have investigated the integration of flight mechanics and aeroelastic control design.

As the aircraft rigid-body motion and the elastic degrees of freedom are highly coupled, with mode shapes and frequencies changing with the flight conditions and loading, a realistic aircraft model has to be generated. Here, linearized integrated flight mechanics and aeroelastics models are generated as outlined in Ref. 4. In addition, a simulation model with nonlinear rigid-body dynamics is used for flight maneuver verification. Model reduction techniques (Ref. 6) are employed to generate separate control design models for the longitudinal motion.

In addition to the sensor information obtained from an inertial platform, accelerometers placed along the aircraft structure are used. Control is based on conventionally available control surfaces for primary flight control, i.e., elevator, rudder and inner and outer ailerons. While in Ref. 2 symmetrically deflected inner ailerons are available as means of direct lift control, here symmetric inner and outer aileron activity is restricted to low authority aeroelastic control purposes.

The flight control system for the longitudinal motion is divided into an outer-loop flight path and attitude control and an inner-loop stability augmentation and aeroelastic control. Emphasis in this paper is put on the inner-loop control, which is assumed to be embedded in an outer-loop structure. This outer-loop controller may be based on the concept of total energy control (TECS, Ref. 7) or may be even more elaborate with various autopilot functions. For this paper, it is assumed that the outer loop produces essentially a reference command for the desired C^* command, where C^* is a combination of the vertical acceleration at the pilot position and pitch rate.

Three alternatives are discussed for the inner-loop control system design. First, a conventional cascaded single-input-single-output (SISO) design is presented, which improves the flying qualities of the aircraft without any active aeroelastic control. The second approach is based on output feedback and does influence the rigid body as well as the aeroelastic dynamic behavior of the aircraft. As a third approach, H_∞ optimization is used to design the controller for the inner loop. This gives a robust design with respect to the different operating points of the aircraft.

2. Modeling the Dynamic Aircraft Behavior

In general, the rigid-body dynamics of an aircraft is described by the equations of motion consisting of 12 nonlinear scalar differential equations with 3 states x,y,z for the position of the aircraft center of mass, 3 states u,v,w for the velocity components in a body-fixed reference coordinate system, 3 states ψ, ϑ, φ, that is, azimuth, pitch, and roll angles to describe the attitude with respect to an Earth-fixed reference coordinate system, 3 states p,q,r, that is roll, pitch, and yaw rates around the body-fixed axes. The controls are the elevator, rudder, and aileron angles and the power setting. A detailed description of these equations is given for instance in Ref. 8.

2.1. Structural Dynamics. The structural dynamics for the static and dynamic deformations of the aircraft is described by linear differential equations, which are generated using the finite-element method (FEM, see e.g. Ref. 9). To arrive at such a model, the structure is assumed to consist of many geometrically simple parts, the finite elements. In every element, a space-dependent displacement function is approximated by a fixed number of interpolation functions, describing the displacement behavior of the element as a function of the displacement z of the discrete nodes. Integrating over the element volume (and the known interpolation functions), the work done by the inertial, elastic stiffness, and external forces is expressed as a function of the nodal displacements and the external nodal point forces. Assembling the results for the individual elements, a system of second-order differential equations for the nodal displacements is obtained,

$$M_{FE}\ddot{\vec{z}} + K_{FE}\vec{z} = P_{FE}\vec{f}. \tag{1}$$

In equation (1), $\vec{z}$ is the vector of nodal displacements and $\vec{f}$ is the vector of external point forces. M_{FE} denotes the mass matrix, K_{FE} the stiffness matrix and P_{FE} the load matrix of the aircraft. For the determination of the static deformation of the structure $\left(\ddot{\vec{z}} = 0\right)$, the algebraic equation $K_{FE}\vec{z} = P_{FE}\vec{f}$ has to be solved. For the dynamic deformation, the solution of the homogeneous differential equation is determined by setting $\vec{z} = \phi e^{\lambda t}$, thereby separating the time-dependent and space-dependent components of the solution and solving the resulting eigenvalue problem,

$$\left[\lambda^2 M_{FE} + K_{FE}\right]\phi = 0. \tag{2}$$

The eigenvectors ϕ_h describe the mode shapes (normal modes) of the undamped structure. For a free-flying aircraft structure, 6 zero eigenvalues, representing the rigid body motion are obtained. The corresponding eigenvectors ϕ_{rh} can be chosen to represent the unit displacements in the direction of the axes of the center-of-mass-based, body-fixed reference frame and the unit rotations about these axes. The eigenvectors ϕ_{eh} (orthogonal to the rigid body motion) associated with the negative eigenvalues describe the elastic deformations of the structure at the fixed center of mass. They are normalized with respect to the mass matrix,

$$\phi_{ei}^T M_{FE}\phi_{ej} = \begin{cases} 0, i \neq j \\ 1, i = j \end{cases} \text{ and therefore } \phi_{ei}^T K_{FE}\phi_{ej} = \begin{cases} 0, i \neq j \\ \lambda_h^2, i = j \end{cases}. \tag{3}$$

The corresponding eigenmotions $e^{i\lambda_h t}$ of the undamped structure are harmonic oscillations with eigenfrequency λ_h. Now, any small arbitrary motion and deformation of the aircraft can be represented (within the resolution of the discretization) by the superposition of the free undamped normal modes,

$$\bar{z}(t) = \sum_{h=1}^{n_{FE}} \phi_h q_h(t) = \Phi \bar{q}(t), \tag{4}$$

with $\Phi = (\Phi_r, \Phi_e)$. Here Φ_r is the matrix of the eigenvectors for the rigid body motion, Φ_e is the matrix of the eigenvectors for the elastic motion, and q is a vector of generalized coordinates.

A good approximation can be achieved by retaining only a small number of modes at the low-frequency end of the set. For the flight-mechanical and aeroelastic analyses addressed in this paper, the aircraft motion can be described with sufficient precision using n_q modes (6 rigid body modes plus up to 60 low-frequency elastic modes for a full aircraft model) up to about 20Hz,

$$\bar{z}(t) \approx \sum_{h=1}^{n_q} \phi_h q_h(t) = \Phi \bar{q}. \tag{5}$$

Inserting the approximation of equation (5) into equation (1) and left multiplying by Φ^T, a compact representation of the aircraft motion and deformation can be achieved using a relatively small number of generalized coordinates in the vector $\bar{q}$,

$$M\ddot{\bar{q}} + \Lambda^2 \bar{q} = P\bar{f}, \quad \bar{z} = \Phi\bar{q}, \quad \dot{\bar{z}} = \Phi\dot{\bar{q}}, \quad \ddot{\bar{z}} = \Phi\ddot{\bar{q}}. \tag{6}$$

Additional vectors, the control modes ϕ_c, describing the unit deflections of the control surfaces, are added to the eigenvector matrix $\Phi = [\Phi_r, \Phi_e, \Phi_c]$. The control surface motion is appended with given spring constants and mass and stiffness matrices in generalized coordinates. The inertial coupling of the control surface motion and elastic deformation is neglected. In a later step, a transfer function representing the actuator dynamics is added.

2.2. Aerodynamic Forces and Moments. The air flow around a flexible aircraft is modeled as an inviscid compressible flow. For the purpose of aeroelastic calculations, the doublet-lattice-method (DLM) for the approximate numeric calculation of the unsteady pressure distribution on harmonically oscillating surfaces in three-dimensional subsonic flow was developed by Albano and Rodden (Ref. 10). In this approach, the surface of the aircraft structure is discretized by means of trapezoidal boxes arranged in columns parallel to the free stream, the so-called panels. The 1/4-chord line of each box is taken to contain a distribution of acceleration potential doublets, expressed as local pressure differences, of uniform but unknown strength. Then, an integral equation for the induced downwash can be solved approximately for individual reduced frequencies $k = \omega c / u_\infty$, with ω as the undamped rigid body or structural frequency, c the wingspan or the mean chord, and u_∞ the free stream velocity.

The resulting forces (normal to the plane of the box) and moments (about the 1/4-line of the box) are obtained by multiplying the pressure difference over each box with the box area. Using the above technique, it is possible to calculate a matrix of influence coefficients $Q_{DLM_{ij}}(k, Ma_\infty)$ that relates the changes in the lifting force at box i to the changes in the induced downwash at box j. This influence coefficient matrix has to be calculated for different Mach numbers and a number of frequencies in the range of interest.

The DLM is well suited to account for the influence of wings and tail planes. The power plants are modeled as annular wings. The influence of the fuselage can be treated approximately. To extend the use of the DLM to the transonic flight regime, the calculated pressure distributions can be calibrated using a nonlinear Euler solution for steady flows.

The result of the aerodynamic force and moment calculations using this method is

$$\vec{f}_a = P_d \left[G^T Q_{DLM}(k, M_\infty) G \right] \vec{z} . \tag{7}$$

Here, P_d is the dynamic pressure, G is a matrix that provides an interpolation between the structural noding and the boxes used for the aerodynamic calculations. Introducing this into equation (1) yields the relation

$$M_{FE} \ddot{\vec{z}} + K_{FE} \vec{z} = P_{FE} T_{fa} P_d \left[G^T Q_{DLM}(k, M_\infty) G \right] \vec{z} , \tag{8}$$

where T_{fa} is a transformation matrix from aerodynamic to body-fixed axes. Transforming equation (8) to the frequency domain using the reduced Laplace variable $\tilde{s} = sc/u_\infty$ and assuming a structural damping matrix B_{FE} gives

$$\left[-(u_\infty/c)^2 M_{FE}\tilde{s}^2 + (u_\infty/c) B_{FE}\tilde{s} + K_{FE} - p_d T_{fa} P_{FE} G^T Q_{DLM}(k, M_\infty) G\right]\hat{z} = 0 \tag{9}$$

with the reduced frequency $k = \mathrm{Im}(\tilde{s})$ and the complex matrix $Q_{DLM}(k, M_\infty)$. Equation (9) is called the flutter equation. The complex eigenvalues $\tilde{s}$ and eigenvectors $\hat{z}$ of the nonlinear eigenvalue problem have to be determined iteratively. Flutter occurs if $\mathrm{Re}(\tilde{s}) \geq 0$ for any eigenvalue $\tilde{s}$. The corresponding eigenvector $\hat{z}$ determines the flutter shape. Flutter calculations are described in Ref. 9.

With the aerodynamic forces available, a steady-state trim solution of the flexible aircraft can be computed. To this end, the differential system

$$(p_d T_{fa} P_{FE}\left[G^T Q_{DLM}(0, M_\infty) G\right] - K_{FE})\bar{z} = M_{FE} T_{fg} \ddot{z}_g \tag{10}$$

must be solved for the deformation vector $\bar{z}$ with given acceleration due to gravity $\ddot{z}_g$ and a transformation matrix T_{fg} from a geodetic system to a body-fixed coordinate system.

Having obtained the deformation vector $\bar{z}$, a coordinate transformation to the stability axes at a particular Mach number is performed. With the definition

$$Q(k, M_\infty) = \Phi^T G^T Q_{DLM}(k, M_\infty) G \Phi, \tag{11}$$

the generalized aerodynamic forces in the new coordinate system can be expressed as

$$\bar{f}_a = P_d Q_o + P_d Q(k, M_\infty)\bar{q}. \tag{12}$$

Here, the term Q_0 represents the steady-state aerodynamic force corresponding to the trim angle of attack at a particular Mach number.

2.3. State Space Description. Aerodynamic forces based on the generalized DLM aerodynamic force coefficient can be evaluated only for harmonic oscillations at given discrete reduced frequencies k. To realize a time domain simulation model, the unsteady aerodynamic forces $\bar{f}_a$ have to be represented in the time domain. This can be achieved first by approximating the tabulated force coefficients by rational functions of the Laplace variable s and then by transforming the resulting transfer functions to the time domain.

The major difficulty associated with rational function approximation is the matching of phase responses dominated by phase lags due to dead times. Dead times occur frequently in unsteady aerodynamic responses, representing for example the transit time from the wing to the tailplane. However, rational transfer functions and continuous-time state-space representations allow no exact representation of dead times and delays. Instead, a large number of additional lag states is required to provide the necessary phase lag. While the lag states are a common feature of all rational function approximations, the number of lag states required for different methods varies considerably. As a large number of lag states means higher model complexity and increased calculation effort, methods requiring a lower number of lag states are preferable for industrial-size problems.

The minimum-state method of Ref. 11 formulates a general rational transfer function matrix,

$$\widetilde{Q}(\widetilde{s}) = A_0 + A_1\widetilde{s} + A_2\widetilde{s}^2 + D(I\widetilde{s} - R)^{-1}E\widetilde{s}\,, \tag{13}$$

in the reduced Laplace variable $\widetilde{s}$ to match the tabulated coefficient matrix on the imaginary axis, that is,

$$Q^f(k) \approx \widetilde{Q}(jk)\,. \tag{14}$$

where $k = \mathrm{Im}(\widetilde{s})$. The diagonal matrix R in equation (13) is used to define the aerodynamic lag states. Usually, roots with absolute values spread within the range of the tabulated reduced frequencies are chosen. The elements of the matrices A_0, A_1, A_2, D, E are determined from a nonlinear weighted least-square solution minimizing, under some constraints, the total weighted least-square approximation error.

Up to 3 constraints for every element of Q^f can be introduced to enforce perfect data fit at specific frequencies (for example at k=0). Weights are used to improve data fitting for selected elements at specific frequencies, or simply for normalizing the tabulated data.

From the transfer function matrix in equation (13), a time domain state-space representation can be derived by changing from $\tilde{s}$ to s (the unscaled Laplace variable) and then applying the inverse Laplace transform. It should be remembered though that both the aerodynamic coefficient matrix Q^f and the set of approximation matrices resulting form the minimum-state method are valid for only a single Mach number and trim condition. Therefore, the simulation of a flight trajectory, when the Mach number changes, requires an interpolation between different sets of matrices. A time-domain representation of the aerodynamic forces, with coefficient matrices A_0, A_1, A_2, D, E (R is assumed constant) scheduled with Mach number is given in equations (15)-(16), where the vector $\bar{x}_a$ representing the aerodynamic lag states is introduced,

$$\bar{f}_a = T_{fa}\Big[p_d Q_0^f(M_\infty) + p_d A_0(M_\infty)\bar{q} + (p_d / u_\infty)A_1(M_\infty)\dot{\bar{q}} + \rho A_2(Ma_\infty)\ddot{\bar{q}} + D(M_\infty)\bar{x}_a\Big], \tag{15}$$

$$\dot{\bar{x}}_a = (u_\infty)^{-1} R x_a + E(M_\infty)\dot{\bar{q}}. \tag{16}$$

For the scheduling, an interpolation scheme based on third-order Hermite polynomials and the (evidently false) assumption of zero tangent at the Mach grids is used.

With an approximation of induced drag in place, it is possible to complete the translational equation of motion in the body-fixed x-direction by adding thrust and ram drag. In a flexible aircraft, the thrust vector moves with the powerplant during vibrations, while the ram drag depends on the local flow condition. After summing up the forces of the powerplants and generalizing, the thrust forces can be described by $\bar{F}_t = \bar{F}_{t0} + T_q\bar{q} + T_{\dot{q}}\dot{\bar{q}} + T_t\bar{u}_t$, where $\bar{F}_{t0}$ denotes the thrust forces at the trim condition, $T_q\bar{q}$ and $T_{\dot{q}}\dot{\bar{q}}$ denote the linearized thrust forces depending on aircraft motion and deformation, and $T_t\bar{u}_t$ denotes the thrust forces due to changes in the throttle position (throttle position vector $\bar{u}_t$).

In stability axes, the coupled flight mechanics and aeroelastic equation can finally be described as

$$M\ddot{\bar{q}} = -H(\dot{\bar{q}}) - 2\zeta_m \Lambda \dot{\bar{q}} - \Lambda^2 \bar{q} + T_{fa}\left[p_d \hat{Q}_0^f(M_\infty) + p_d A_0(M_\infty)\bar{q}\right]$$
$$+ T_{fa}\left[(p_d / u_\infty) A_1(M_\infty)\dot{\bar{q}} + \rho A_2(M_\infty)\ddot{\bar{q}} + D(M_\infty)\bar{x}_a\right]$$
$$+ B\bar{u}_c + M T_{fg}\bar{g} + \bar{F}_t(\bar{q}, \dot{\bar{q}}, \bar{u}_t), \qquad (17a)$$

$$\dot{\bar{x}}_a = (u_\infty^{-1}) R \bar{x}_a + E(M_\infty)\ddot{\bar{q}}\,. \qquad (17b)$$

Here, M is the generalized mass matrix, H is a generalized matrix that contains the Coriolis terms due to the moving coordinate system, Λ is the generalized stiffness matrix, B is the input matrix obtained from a partitioning of the doublet lattice matrix, $\bar{u}_c$ is the control input consisting of the inner ailerons and outer ailerons, rudder, and elevator, T_{fg} is a transformation matrix from geodetic coordinates to stability axes, $\bar{g}$ is the gravitational acceleration, and $\bar{u}_t$ has been described above. For a detailed derivation of these equations, see e.g. Ref. 4.

3. Analysis of the Aircraft Dynamics

In this paper, a heavy four-engine transport aircraft is chosen as an example. The cruise condition is set at a speed of Mach 0.86 and an altitude of 30000ft. Three additional flight and load conditions are chosen for detailed analysis, see Table 1. Although they represent only a small part of the flight envelope, they allow us to develop an understanding of the basic phenomena related to changes in the flight condition (speed, altitude, and correspondingly Mach number and dynamic pressure) and the load condition (tank loading, changing mass, moments of inertia, and c.g. position). Flight condition 1 represents the cruise condition and is chosen

Table 1. Four reference flight conditions for the equations of motion:

Case No.	Mach No.	Altitude	Inner wing tank	Outer wing tank	Trim tank
(1) IE0-86	0.86	30000 ft	100%	100%	0%
(2) IE0-70	0.70	25000 ft	100%	100%	0%
(3) IET-70	0.70	25000 ft	100%	100%	100%
(4) 000-70	0.70	25000 ft	1%	1%	0%

as the control design flight condition. It is the most challenging from an aeroelastic point of view. Flight conditions 2 and 3 are encountered during climb and flight condition 4 is encountered during descent.

In order to analyze the dynamic response of the flexible structure as a function of various input signals, the frequency responses are investigated. The analysis is based on individual single-input-single-output (SISO) transfer functions. For the example aircraft, measurements at the cockpit, at the center and aft fuselage positions, on the wing, and at the engines are chosen. The main results are discussed using only a limited number of transfer functions from the longitudinal motion.

Figure 1 shows the transfer functions from elevator deflection on cockpit vertical acceleration; Fig. 2 shows the transfer function from symmetric inner aileron deflection on midwing vertical acceleration. The rigid-body modes (phugoid and short period mode) can be identified easily in the low-frequency domain. In the 1-10 Hz frequency range however, the aircraft response is dominated by weakly damped elastic modes, especially wing bending, engine and fuselage modes. Higher-frequency

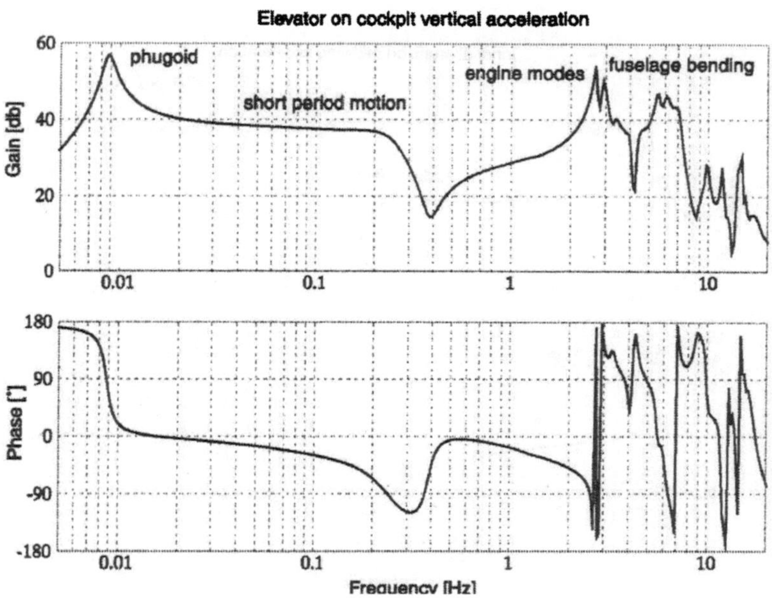

Fig. 1. Transfer functions from elevator to cockpit acceleration for the longitudinal motion.

modes also have considerable influence, but are more difficult to identify on the basis of these transfer functions. Comparing the two transfer functions of the longitudinal motion, it can be noticed that the first (symmetric) wing bending mode is not perceived in the cockpit but dominates the response at the wing, while the outer engine vertical vibration mode (coupled with the wing torsion) interacts with the first fuselage bending mode and can be measured over all of the aircraft. For the given configuration, the outer engine vertical vibration mode (together with the fuselage bending) is the most critical with respect to flutter.

From the gain amplitudes in the aeroelastic frequency range, it can be concluded that the control bandwidth for a flight mechanics stability augmentation system that does not affect aeroelastics must not exceed 1Hz. This restriction limits severely the achievable handling qualities. It is less severe for an integrated flight and aeroelastic control law. But even then, a steep descent to cut off those elastic modes that are to remain unaffected by the control law is required.

The load distribution (fuel and payload) influences strongly the dynamic behavior of the elastic structure and consequently the aeroelastic

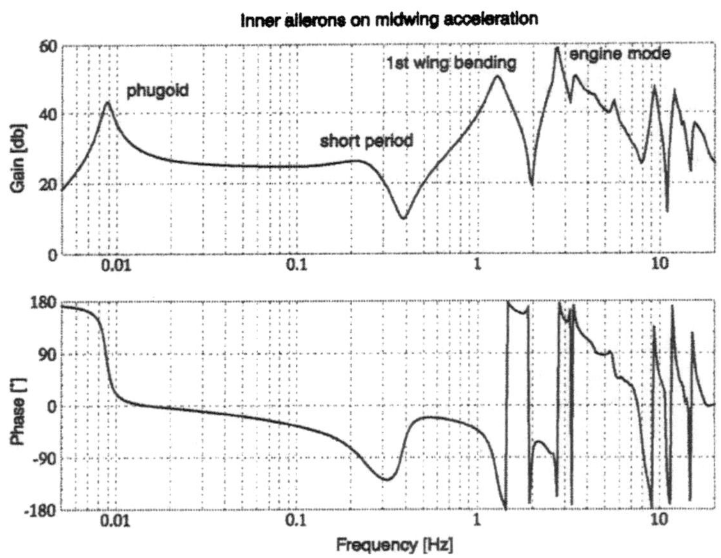

Fig. 2. Transfer function from inner ailerons to acceleration at the midwing position for the longitudinal motion.

coupling. Figure 3 shows the frequency response for three different load conditions (IE0-70, IET-70 and 000-70) at constant flight condition. While changes in the elastic mode shapes and frequencies are not unexpected (wing bending frequency should increase with fuel consumption), strong changes in the damping (maximum amplitudes) and phase response are also observed. As expected, the low-fuel configuration (000-70) turns out to be the least critical with respect to aeroelastics. With mass and moments of inertia reduced, comparatively more control power is available. Therefore, the analysis described in this paper is concentrated on the high-load cases.

Although frequency responses are the preferred means of analysis, time responses are of interest for an assessment of the aircraft handling qualities and for developing a physical understanding of the accelerations and the level of vibration experienced by the pilot and the passengers. The input signals used for the simulations shown subsequently have been designed not to contain frequencies beyond about 5 Hz and could be reproduced by a pilot. Therefore, the curves represent the basic low-frequency response felt by the pilot and the passengers and targeted by the flight and aeroelastic control effort.

Figure 4 shows the time response to the same elevator pulse for different flight conditions. It can be seen that the amplitude varies differently with the flight condition for the pitch rate and cockpit vertical

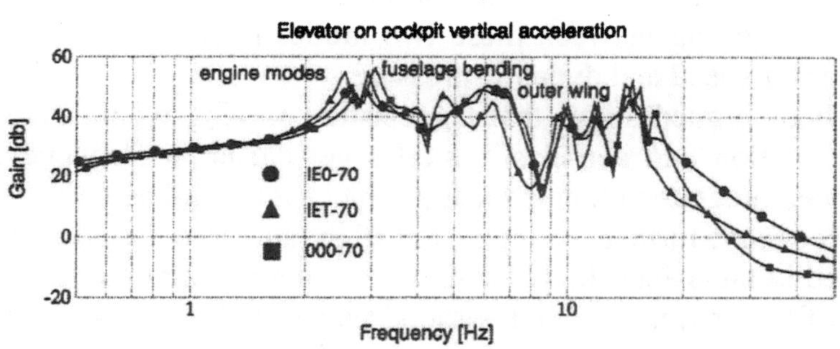

Fig. 3. Influence of load conditions on the aircraft dynamics.

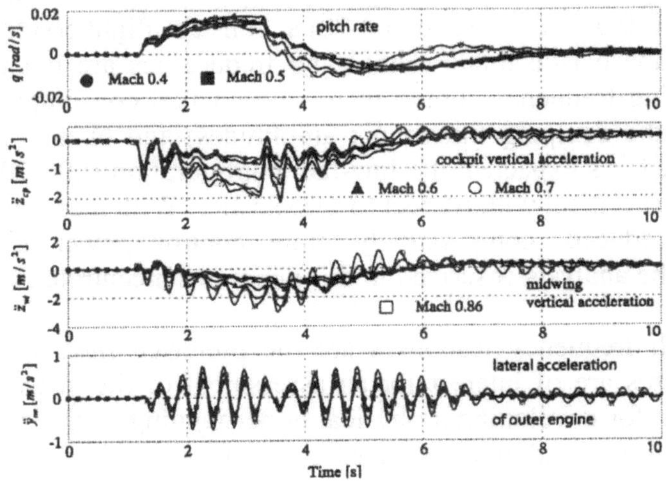

Fig. 4. Time response at various positions of the aircraft due to an elevator impulse at about 1 sec into the flight.

acceleration. This is due to the fact that, at lower speed, a larger angle of attack is required to generate the same amount of lift and consequently vertical acceleration. This relationship should be kept in mind for control design. The acceleration response at different positions of the structure is dominated by the fuselage bending and outer engine mode vibrations. As said before, coupling between these two modes is strong and intensifies with increasing speed and dynamic pressure.

It has been argued above that aeroelastic coupling would increase as rigid-body motion and aeroelastic modes get closer in frequency. As a consequence, integrated models for flight mechanics and aeroelastics were deemed necessary; further, an integrated flight and aeroelastic control law is envisaged in this paper. In this context, it is interesting to investigate the influence of the frequency neighborhood between rigid-body motion and aeroelastic modes on aeroelastic coupling. To that end, two state-space models for the example aircraft in cruise flight have been generated with stiffness changed to 50% and 200% of the nominal value. Figure 5 compares the frequency responses of these models to the response of the nominal model.

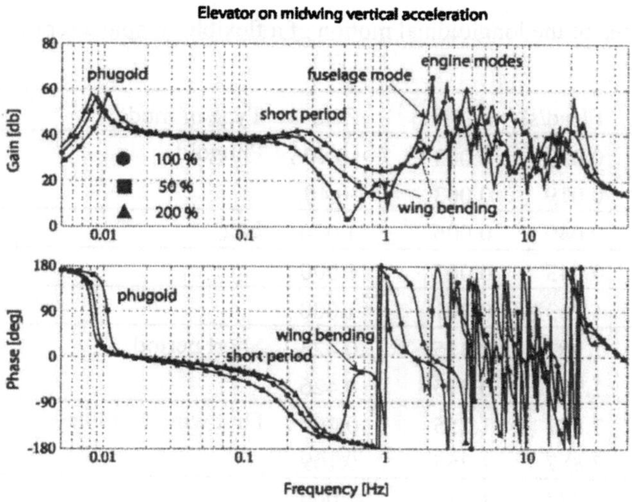

Fig. 5. Aeroelastic coupling for various (assumed) degrees of stiffness of the aircraft.

It can be seen that aeroelastic coupling indeed increases with decreasing stiffness and vice versa. The rigid-body mode frequencies are also strongly affected by the elastic stiffness, with phugoid frequency decreasing and short-period frequency increasing with the stiffness. As could be expected, the frequency response of the stiffer aircraft tends toward the response of the models with fewer or no elastic modes. For the aircraft model with reduced stiffness, damping of the fuselage bending and outer engine vertical vibration modes is lower than for the nominal model and coupling between short-period motion and wing bending is significant (see the change in the phase response).

Table 2 contains the modes of a reference aircraft for $n_q = 7$. One can see easily that the phugoid is unstable. Due to the long period of approximately five minutes, however, this would be controllable easily by a pilot. The short period mode is rather well damped, its frequency is about one fourth of the frequency of the lowest elastic mode, which is the first wing bending mode. All elastic modes are close to the imaginary axis, that is, they have low damping. Due to the lag states, which are used to approximate the aeroelastic phase lags, the pole positions loose some significance, as the frequency and damping of the aeroelastic modes are not determined uniquely by the dominant (2nd order) poles.

Table 2. Modes of the longitudinal motion of a flexible sample aircraft for $n_q = 7$.

Poles	f [rad/s]	ω_0 [Hz]	Zeta	Type of mode	See Fig.9
0.01 + 0.02i	0.019	0.003	-0.406	Phugoid	
0.01 - 0.02i	0.019	0.003	-0.406		
-0.040	0.040	0.006	1.000	Lag	
-0.043	0.043	0.007	1.000		
-1.26	1.263	0.201	1.000		
-0.84 + 1.79i	1.981	0.315	0.426	Short period	(8)
-0.84 - 1.79i	1.981	0.315	0.426		
-0.86 + 7.84i	7.887	1.255	0.109	First wing bending	(7)
-0.86 - 7.84i	7.887	1.255	0.109		
-8.87	8.874	1.412	1.000	Lag	
-0.23 + 16.01i	16.010	2.548	0.015	Inner engine lateral	(5)
-0.23 - 16.01i	16.010	2.548	0.015		
-1.07 + 16.29i	16.325	2.598	0.066	Outer engine vertical	(6)
-1.07 - 16.29i	16.325	2.598	0.066		
-0.42 + 16.87i	16.871	2.685	0.025	Second fuselage bending	(4)
-0.42 - 16.87i	16.871	2.685	0.025		
-0.63 + 18.67i	18.678	2.973	0.034	Outer engine lateral	(3)
-0.63 - 18.67i	18.678	2.973	0.034		
-0.50 + 20.71i	20.711	3.296	0.024	Wing bending axial	(2)
-0.50 - 20.71i	20.711	3.296	0.024		
-23.06	23.057	3.670	1.000	Lag	
-0.77 + 25.12i	25.133	4.000	0.031	Second wing bending	(1)
-0.77 - 25.12i	25.133	4.000	0.031		
-48.79	48.791	7.766	1.000	Lag	
-85.55	85.555	13.617	1.000		
-134.12	134.118	21.346	1.000		

4. Controller Design

As mentioned above, the controller design of the outer loop is not considered. This controller may be a flight path angle controller, or a speed hold controller, or an altitude hold controller. Here, a C^* command is the reference command for the inner loop with

$$C* = \frac{1}{g}(-\ddot{z}_{cg} + \dot{q}_{cg} x_p), \tag{18}$$

where x_p is position of the pilot with respect to the center of gravity; the other variables are explained below. This function is commonly used to specify the flying qualities of an aircraft. The time history of $C*$ is supposed to be between the lower and upper time history bounds. For the control system design, the function is computed in the feedback loop in Fig. 6 and is compared to a commanded value $C*_c$.

The design goals are: (i) to stabilize the phugoid and to increase the damping of the short period mode; (ii) to reduce structural vibrations as well as to increase passenger comfort; (iii) to increase the damping of the aeroelastic modes up to about four Hz. In addition, it is required that the closed-loop system is robust with respect to various operating points, if possible without scheduling the controller.

Figure 6 shows a possible architecture for the inner-loop control. At the core of the model is the state space system describing the linearized equations of motion at a particular operating point, here operating point 1, see Table1. The state vector is defined as

$$\vec{x} = (z, \vartheta, q_{s1} \dots q_{snq}, u, w, q, \dot{q}_{s1} \dots \dot{q}_{snq}, x_{a1} \dots x_{anq})^T. \tag{19}$$

Here, z, ϑ, u, w, q are associated with the rigid-body motion, $q_{s1} \dots q_{snq}$, $\dot{q}_{s1} \dots \dot{q}_{snq}$ with the structural motion, and $x_{a1} \dots x_{anq}$ are the lag

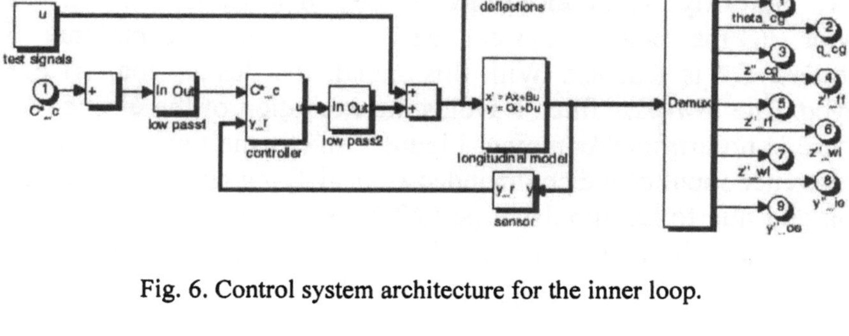

Fig. 6. Control system architecture for the inner loop.

states. The aircraft is controlled by

$$\vec{u} = (\eta, \xi_{si}, \xi_{so})^T, \tag{20}$$

where the components are the elevator and the inner and outer symmetric ailerons, which can be used with low authority for longitudinal control. It is assumed that the following measurements are available:

$$\vec{y} = (\vartheta_{cg}, q_{cg}, \ddot{z}_{cg}, \ddot{z}_{ff}, \ddot{z}_{rf}, \ddot{z}_{wi}, \ddot{z}_{wl}, \ddot{y}_{ie}, \ddot{y}_{oe})^T, \tag{21}$$

where the first two components are attitude, attitude rate at the center of mass, the third component is the acceleration at the center of mass. In the order of appearance, the following components are the vertical accelerations at the forward fuselage, at the rear fuselage, the midwing acceleration at the wings, the acceleration at the winglets, and the lateral accelerations of the inner and outer engines. These measurements can be used for the control system design.

The matrices *A, B, C, D* in Fig. 6 are the results of the linearization process. Between the controller, there is a low-pass filter which filters out any higher-frequency signals which the controller might produce; in front of the controller, there is an additional low-pass filter which filters out any high frequency commands in C^*_c. Below the plant dynamics box, there is a measurement box which selects those output signals to be fedback to the controller. On top of the figure, the actuator signals are recorded in the simulation of the closed-loop system (CLS); on the right side of the figure, the output signals are recorded. The box entitled "test signals" generates perturbations while simulating the CLS.

4.1. Stability Augmentation. If one disregards the aeroelastic behavior in the control system design, like in conventional SAS controllers, C^* is fedback. With this signal, the design goal (i) can be achieved. The low-pass filter 1 avoids the excitation of the elastic modes, but there is no artificial damping. Figure 7 shows the time responses due to a reference input of the commanded C^* satisfying certain flying quality criteria and due to an impulsive perturbation of the inner ailerons of 3.5 degrees magnitude after four seconds. Considerable vibrations with low

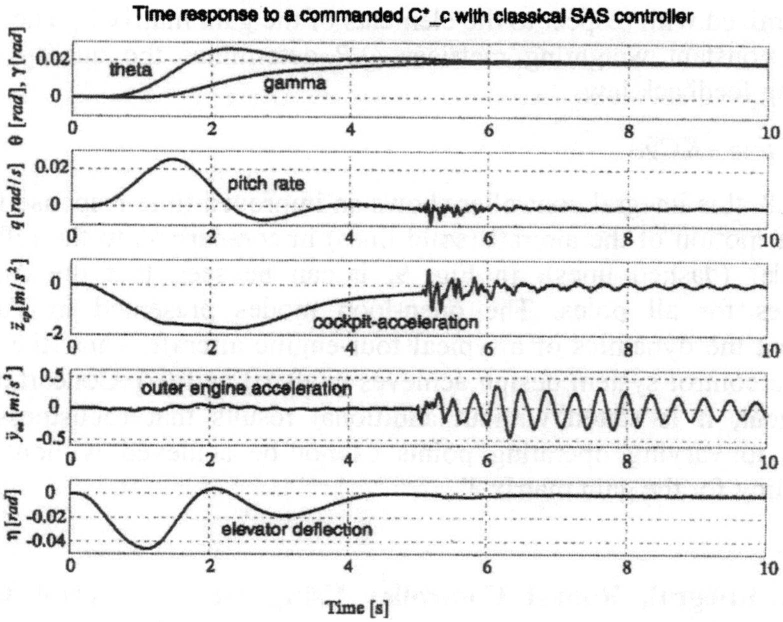

Fig. 7. Time response of a flexible aircraft due to a unit step attitude command and a (gust) perturbation at the inner ailerons at 5 sec into the flight.

damping in the cockpit and at the outer engines are observed. In practice, this would not be tolerable and a separate aeroelastic controller could be designed which improves the damping of the elastic modes. This simple SAS controller is used in the sequel as a reference in order to quantify the improvements which advanced control design methods may offer.

4.2. Integral Controller Using Output Feedback. In addition to the elevator, symmetric inner and outer ailerons are used as actuators. Furthermore, all or some of the available output signals are fedback. Then, the control design problems is formulated as a quadratic output feedback control problem in which the cost functional

$$J = \int_0^\infty \left(\vec{x}^T Q \vec{x} + \vec{u}^T R \vec{u}\right) dt \tag{22}$$

is minimized with respect to the elements of the gain matrix K. The choice of the constant weighting matrices Q,R determines the quality of the resulting feedback law,

$$\vec{u} = -KC\vec{y}. \tag{23}$$

In Fig. 8, this integral controller shows an improved time response for the flexible motion of the aircraft (solid lines) in comparison to the reference controller (dashed lines). In Fig. 9, it can be seen that the damping increases for all poles. The open-loop modes presented in Table 2 represent the dynamics of a typical four-engine aircraft. Thus, the multi-variable control system design achieves goals (i) and (ii). Concerning the third goal, it is stated without additional results that robustness with respect to varying operating points cannot be achieved without some scheduling for the gain matrix K.

4.3. Integral, Robust Controller Using the H_∞ Control Design Method. The advantages of the previous design methods are a clear structure with a unique assignment of dynamic elements (filters,

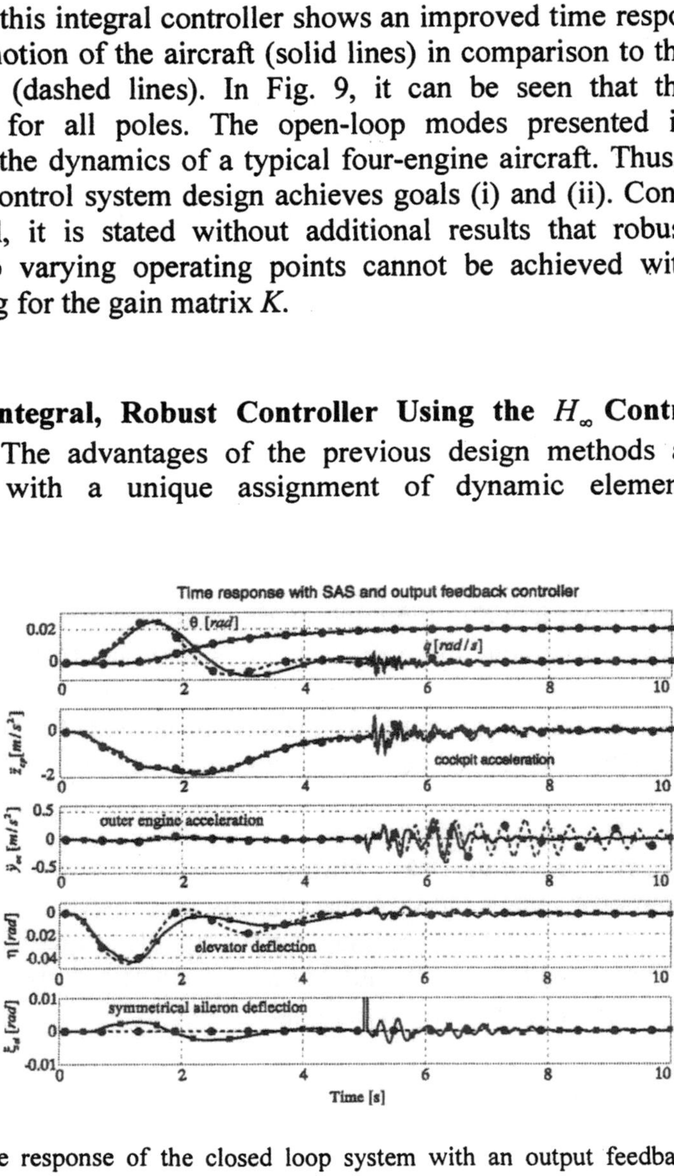

Fig. 8. Time response of the closed loop system with an output feedback controller (reference controller dashed lines, output feedback solid lines).

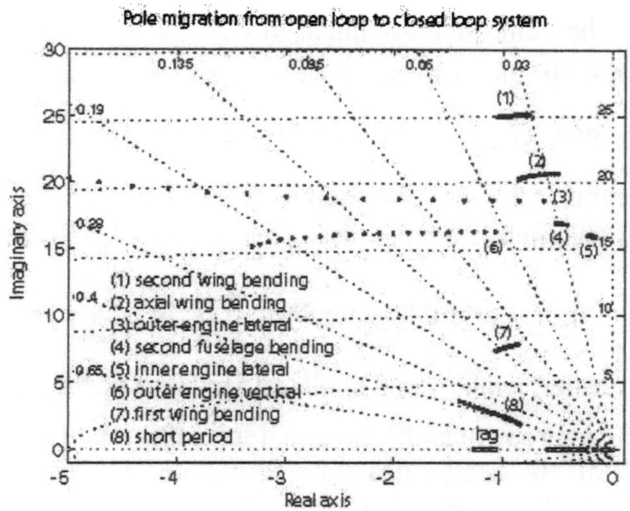

Fig. 9. Migration of open-loop poles using output feedback control.

integrators, sensors) to particular tasks. This makes it possible to define structured redundancy concepts for those cases where for instance sensors or actuators degrade. The disadvantage of these methods is their lack of robustness.

To compensate the deficiency observed in the output feedback design, the H_∞ design method considers modifications of the nominal plant in the design process; that is, error models for various dynamic components of the system are defined and considered in the design process. In addition, nonmeasured states are estimated through an observer. The design goals are defined in terms of the H_∞ norm of particular transfer functions of the closed-loop system. This norm is a metric of all gains as a function of the frequency. For the transfer function G_{CC_c} from C^*_c to C^*, for instance, one could demand that the closed-loop system should perform like a second-order system with the transfer function G_{ideal}. Then, the requirement in terms of the H_∞ controller is formulated as

$$\left\| G_{CC_C} G_{ideal}^{-1} \right\|_\infty = 1 \quad \text{or} \quad \left\| G_{CC_C} G_{ideal}^{-1} - I \right\|_\infty = \min , \tag{24}$$

with respect to the controller contained in G_{CC_C}. I is 1 in the scalar case and an appropriately dimensioned identity matrix in a multivariable case. The transfer matrix G_{ideal}^{-1} can be viewed as a frequency dependent weighting function. Alternatively, should the influence of a gust with gust velocity w_g on the pitch rate q be minimized, then the criterion should be

$$\left\| G_{qw_g} \right\|_\infty = \min, \quad \text{or more generally,} \quad \left\| G_{qw_g} W_{qw_g} \right\|_\infty = \min, \tag{25}$$

with W_{qw_g} a specified weight. In a similar way, the modeling errors can be formulated. The approach can be extended to multivariable problems; see e.g. Ref. 12. If all the design goals are formulated in this way, then the design task consists of finding a controller $K(s)$, s being the Laplace variable, which minimizes the infinity norm of a transfer matrix G_{zw} describing the influence of the external inputs on the external output z. This approach has been used here and details about the design procedure are given in Ref. 4. Figure 10 shows the time histories for the same

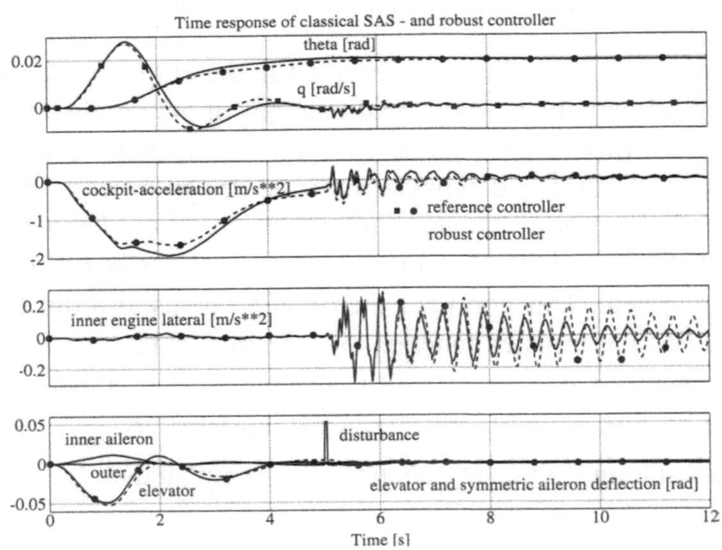

Fig. 10. Time response of the closed-loop system with the H_∞ controller (reference controller dashed lines, robust controller solid lines). Empty symbols refer to the robust controller.

variables as presented in Fig. 8. It can be observed that the robust controller shows actuator activity at the inner and outer ailerons, which the reference controller did not have. Figure 11 shows the pole migration. It can be observed that damping is increased for all modes. It is rather difficult to increase the damping of the engine modes, like in the output feedback controller design. Figure 12 (in two parts) demonstrates that the controller is robust indeed. Here, the same controller is used for simulating unit step responses in C^* with a perturbation after 4 sec at the inner ailerons. The variables shown are defined in equation (20). γ is the flight path inclination. The aircraft response is quite similar for all flight conditions shown and the elastic mode damping is satisfactory.

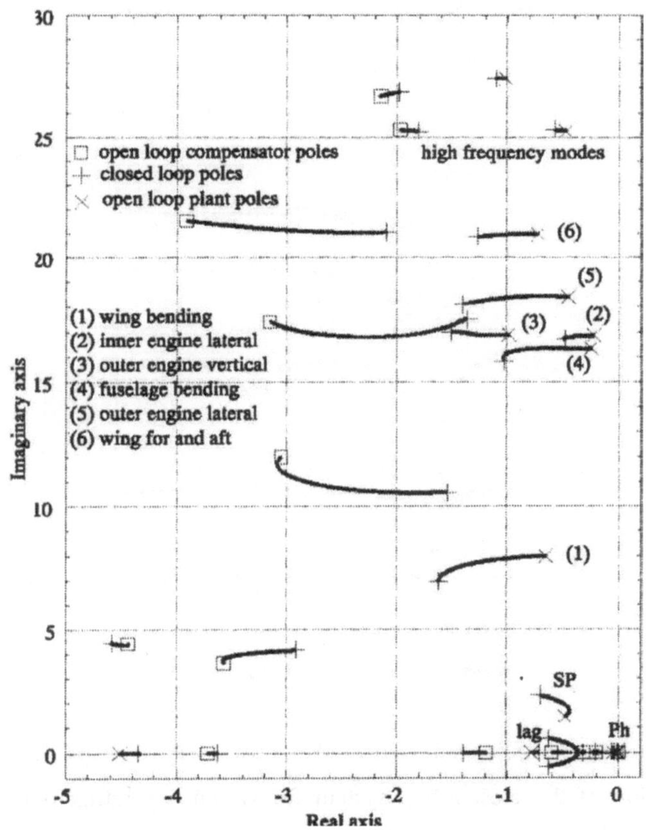

Fig. 11. Migration of open-loop poles using the robust controller.

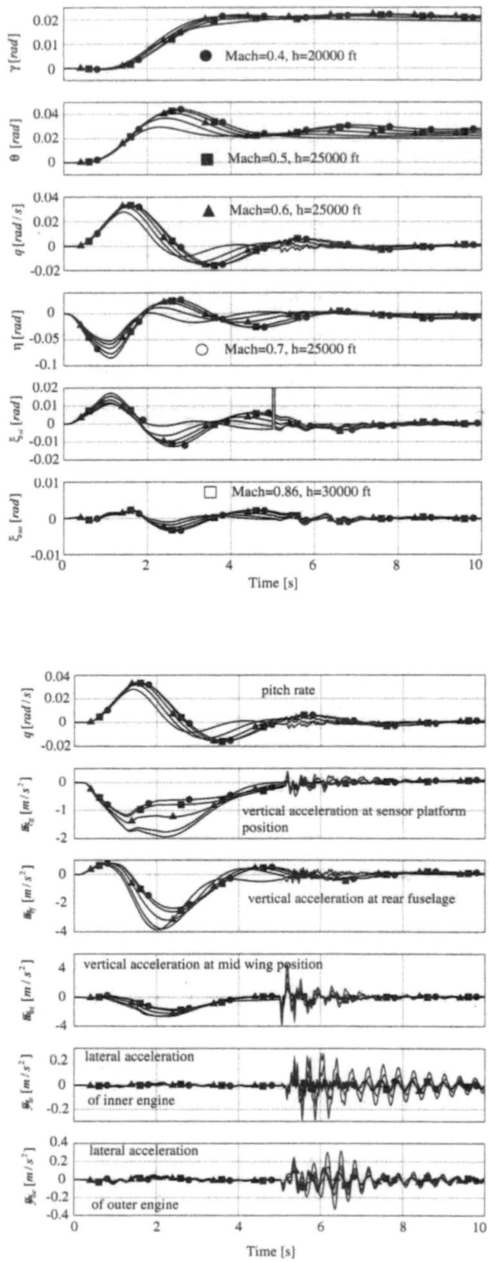

Fig. 12. Simulation of the closed-loop system for various operating points using the H_∞ controller.

5. Conclusions

Based on an integral model describing the dynamic behavior of the rigid motion as well as the elastic motion of a flexible aircraft, it has been shown that an integral controller can achieve desired flying qualities as well as dampen the elastic vibrations considerably. This is achieved by feeding back to the control system not only pitch attitude, pitch rate, and vertical acceleration at the center of gravity, but in addition, various accelerations measured at certain positions of the aircraft. The least damped eigenmode is a symmetric vibration of both the outer engines, in the y-direction of the lateral aircraft axis, which can only be improved marginally through the control system.

References

1. SEYFFARTH, K., et al., *Comfort in Turbulence for a Large Civil Transport Aircraft*, Proceedings of the International Forum on Aeroelasticity and Structural Dynamics, Strasbourg, France, 1993.

2. SCHULER, J., *Flugregelung und aktive Schwingungsdämpfung für flexible Großraumflugzeuge*, Dissertation, Universität Stuttgart, Stuttgart, Germany, 1997.

3. KUBICA, F., and LIVET, T., *Flight Control Law Synthesis for a Flexible Aircraft*, Proceedings of the AIAA Guidance, Navigation and Control Conference, Scottsdale, Arizona, Paper AIAA 94 - 3630, pp. 775-783, 1994.

4. HANEL, M., *Robust Flight and Aeroelastic Control System Design for a Large Transport Aircraft,* Dissertation, University of Stuttgart, Germany, 2000.

5. TEUFEL, P., HANEL, M., and WELL, K. H., *Integrated Flight Mechanics and Aeroelastic Modelling and Control of a Flexible Aircraft Considering Multidimensional Gust Input*, NATO Research and Technology Organization (RTO), Specialist Meeting on Structural Aspects of Flexible Aircraft Control, Ottawa, Canada, 1999.

6. MOORE, B., *Principal Component Analysis in Linear Systems: Controllability, Observability and Model Reduction*, IEEE Transactions on Automatic Control, Vol. 26. No. 1, pp. 17-32, 1981.

7. LAMBREGTS, A., *Vertical Flight Path and Speed Control Autopilot Design Using Total Energy Principles*, Paper AIAA 83-2239, 1983.

8. STEVENS, B. L., and LEWIS, F. L., *Aircraft Control and Simulation*, John Wiley and Sons, New York, NY, 1992.

9. DOWELL, E. H., et al., *A Modern Course in Aeroelasticity*, Kluwer Academic Publishers, Dordrecht, Holland, 1995.

10. ALBANO, E., and RODDEN W., *A Doublet-Lattice Method for Calculating Lift Distributions on Oscillating Surfaces in Subsonic Flows*, AIAA Journal, Vol. 7, No. 2, pp. 279 – 285, 1969.

11. KARPEL, M., and STRUL, E., *Minimum-State Unsteady Aerodynamic Approximations with Flexible Constraints*, Journal of Aircraft, Vol, 33, No. 6, pp. 1190-1196, 1996.

12. DOYLE, J. C., GLOVER, K., KHARGONEKAR, P., and FRANCIS, B., *State-Space Solutions to Standard and Control Problems*, IEEE Transactions on Automatic Control, Vol. 34, No. 8, pp. 831-847, 1989.

General Index

Subject Index

MIX
Papier aus verantwortungsvollen Quellen
Paper from responsible sources
FSC® C105338

If you have any concerns about our products,
you can contact us on
ProductSafety@springernature.com

In case Publisher is established outside the EU,
the EU authorized representative is:
Springer Nature Customer Service Center GmbH
Europaplatz 3, 69115 Heidelberg, Germany

Printed by Libri Plureos GmbH
in Hamburg, Germany